Water Resources Evaluation Methods and Techniques

Water Resources Evaluation Methods and Techniques

A. Ganesh

SATISH SERIAL PUBLISHING HOUSE
SSPH
403, Express Tower, Commercial Complex, Azadpur, Delhi - 110033 (INDIA)
Phone : 011-27672852 Fax : 91-11-27672046
E-mail : info@satishserial.com, hkjain1975@yahoo.com
Website : www.satishserial.com

Published by :

SATISH SERIAL PUBLISHING HOUSE
403, Express Tower, Commercial Complex, Azadpur, Delhi - 110033 (INDIA)
Phone : 011-27672852 Fax : 91-11-27672046
E-mail : info@satishserial.com, hkjain1975@yahoo.com

ISBN 81-89304-21-6

Printed at
SALASAR IMAGING SYSTEMS, DELHI-35

Preface

Water is the most sought resource for human use. Owing to rapid population growth and urbanization as well as rising needs of power generation and modern industry, water has been a major concern, as could be seen from the recent attempts in international cooperation in conserving and utilizing the global water resources. Since water is a component of river basin, its role in river basin research cannot be over emphasized for without it a comprehensive appraisal of development potential of a basin can never be achieved. The present study provides various methods and techniques for evaluation of water resources in the Upper Vaigai basin, South India. The Upper Vaigai basin, located in a rain-shadow region of Theni and Dindigul districts of Tamil Nadu, is one of the beneficiaries of the west flowing river Periyar. This sort of water resource studies are essential to help to modernize water use and find out measures in formulating efficient water use policies.

The first chapter unfolds the subject matter and procedure of the study. The second chapter gives a geographical background of the Upper Vaigai basin. Evaluation of surface and ground water resources along with climatic water balance are analyzed in the third chapter. The fourth chapter examines the development and utilization of surface and ground water resources and provides an insight into the problems and prospects of water resources. The final chapter summarizes the research work and outlines the future course of study.

I express my deep sense of gratitude and indebtedness to late Dr. Mrs. Vasantha Vishwanath, M.A., M.Litt., Ph.D., former Professor of Geography, University of Madras, Chennai under whose guidance and supervision this investigation was carried out. But for her inspiration, encouragement, constructive criticism and valuable suggestions this work could not have been completed to take the present form. My profound gratitude and indebtedness to late Dr. A. Ramesh, M.Sc., Ph.D., M.I.S., former Professor and Head, Department of Geography, University of Madras, Chennai for providing me an opportunity to work in the UGC sponsored research project on 'Planning and Development of Land and Water Resources in the Vaigai - Vaippar Basins, Tamil Nadu' and also for providing me excellent facilities and encouragement to complete my research work. This research work is an offshoot of the above mentioned projects. I am particularly indebted to Prof. S. Subbiah, M.Sc., M.Phil., Ph.D., Editor, Indian Geographical Society, University of Madras, Chennai for his unfailing encouragement, valuable criticism, constructive suggestions and meticulous care throughout my research work.

My thanks to the Regional Director, India Meteorological Department; the Chief Engineer, Ground Water Division and the Chief Engineer, Irrigation Division of the Public works Department, the Director of Statistics of the Government of Tamil Nadu and the Chairman Tamil Nadu Water Supply and Drainage Board, Government of Tamil Nadu for providing me the necessary data, maps and relevant information.

I extend my deepest appreciation to my parents and family members for their continual encouragement and sacrifice for completing this research work.

A. Ganesh

Content

List of Tables

List of Illustration

CHAPTER 1

WATER RESOURCES EVALUATION METHODS AND TECHNIQUES

1.1 INTRODUCTION

Water is indispensable for the existence and survival of life on the Earth. The rapid growth of population together with extension of irrigated agriculture, urbanization and industrial development makes a stressing demand on the quantity and quality of this natural resource and thus it becomes a precious resource. However, the total quantity of water available on the Earth is virtually constant. The rapidly growing demand on the constant supply naturally requires a thorough analysis on its quantity, pattern and use. The analysis will facilitate to have the much needed assessment of adequacy of surface and subsurface water for use in future, thereby laying down directions for a rational and efficient use of this precious resource (UNESCO and WMO, 1981).

Water availability and flow depends largely on the rainfall which varies seasonally and regionally. This temporal and spatial variability necessitates the storage by conserving water in years of excess flow so as to some extent mitigate the deficiency in the years of poor flow, and this is a planned use of available water. This sort of storage and a rational and efficient management of water resources become a pressing need especially in areas drained by non-perennial rivers. A geographical analysis of river basins in order to help in the control and use of water for multiple human purposes therefore becomes imperative (Weber and Hofschmictt, 1963).

India has a geographical area of 3.28 million km^2 with a mean annual precipitation of 110 cm. As in the case of monsoon regions of the world nearly 90 to 95 per cent of this annual rainfall is restricted to three of four months of the year and the rest of the period has to rely on the available stored-water which is the excess from the three or four months rainfall. In addition to this, there is an uneven regional distribution of rainfall as well. The total quantum of surface and groundwater in India has been estimated to be 179 million hectare metre (ha.m) (Rajan, 1978). It is assumed that surface water of above 67 million ha. m and groundwater of about 26.5 million ha. m could be economically tapped. The Five Year Plans accorded the right emphasis to programmes of water resources development. Up to 1951 only 9.3 million ha. m was utilized to irrigate about 26.2 million ha. It is envisaged that during the Sixth Five Year Plan, utilization was 43 million ha. m. It is expected that about 107 million ha of land may eventually be brought under irrigation by utilizing the 93.5 million ha. m of exploitable water. The Government of India has drawn up a programme for increasing the area under irrigation from 56.7 ha. m to 113 ha. m by 2020 A.D.

The purpose of water resources development is to achieve maximum economic and social returns. In the application of knowledge to water resources development, a growing consensus emerges as to what constitutes a proper assessment of such development. It forms a vital link between exploration on one hand and development and management on the other. The assessment and exploration phases eventually culminate in developmental activities through careful planning and administration. Effective use of water resources depend on accurate knowledge of water availability, its spatial and temporal distributions, and water requirements with special reference to place and time of requirement by the users (Thiruvengadachari et al., 1978).

1.2 DRAINAGE BASIN AS A BASIS FOR THE STUDY OF WATER RESOURCES DEVELOPMENT

A drainage basin is a well-defined hydrological unit. In a variety of ways the drainage basin has formed a framework for human activity. Indeed, the idea of the drainage basin as an appropriate areal unit for the organization of human activity and for regional planning has only recently been revived with the recognition of a system in which soil and vegetation cover as well as hydrological balance are involved, and which needs integrated plans and policies to deal with problems posed by flood control, sedimentation, soil erosion, hydro-electric power generation, navigation and even nature conservation and stream pollution (Smith, 1969).

Basin-wise water resource studies are an essential pre-requisite for any type of development planning. With the present approach to rural development through micro-level planning by the governments, both at the Central and the State levels, a thorough study on the availability and the pattern of utilization of water resources at the basin level is of utmost importance. These studies are essential to help modernize the water-use policies. Geographical analysis will assist the proper planning of water resources. The control and use of water for the varied uses of man, in river basins as a whole or in the major sub-basins is important. This concept has been stated by the United Nation's Panel on Integrated River Development (United Nations, 1958). New dimensions have been added with the realization of the importance of measures like soil reclamation, flood control and afforestation which go to make river valley development project truly multipurpose in character (Dayal, 1971).

In the light of international and national trends in the field of water resource studies, an attempt is here made to assess the surface and ground water availability and utilization of water in a basin, and the problems and prospects of the people in the existing environments. This sort of micro-geographical study of late is assuming great importance in India.

1.3 STUDIES ON WATER RESOURES: A REVIEW

It is truism that the distribution of water and its availability are closely associated with the development of human society. Archaeological and documentary discoveries emphasize the significant part played by the location and magnitude of water supplies in the lives of the people of the Old Testament, the ancient Egyptians and later of the Greeks and Romans. There has been a continued search in human history for an explanation of springs, stream flow and the occurrence and movement of groundwater.

Aristotle (384-322 B.C) explained the mechanics of precipitation. Virtuvius, three centuries later, believed in the pluvial origin of springs. Da Vinci (1452-1519), though having had confused ideas about the hydrological cycle, had a much better understanding of the principles of flow in open channels. Palissy (1510-1590) stated categorically that rainfall was the only source of springs and rivers (Biswa, 1969).

It was not until the later half of the seventeenth century, however, that plausible theories about the hydrological cycle, based on experimental evidences, were advanced. The greatest advances came largely through the work of three men, Pierrek Perrault and Edame' Mariotte. Their work on the Seine drainage basin demonstrated that rainfall was more than adequate to account for river flow. Perrault, Mariotte, and Halley undertook hydrological research of the modern scientific type and they may well be regarded as the pioneers of hydrology (Jones, 1963).

After a modest consolidation of the subject during the eighteenth century, there was a remarkably rapid growth of knowledge in hydrology during the nineteenth century, which saw the beginnings of systematic river flow measurements (Biswas, 1969). Groundwater hydrology also developed substantially during this period. For instance, William Smith in 1802 was the first person to apply stratigraphic principles directly to groundwater exploitation (Smith, 1972) in 1856 Darcy laid the foundation of groundwater theory in his report on the public water supply system for Dijon (Biswas, 1969).

The nineteenth century also saw the publication of the first textbook in hydrology. This was Nathaniel Beardmove's Manual of Hydrology published in 1862 which was itself a revision of an earlier work of 1850, Hydraulic Tables. In 1904 Daniel W. Mead, of the University of Wisconsin, Published his note on Hydrology as the first American text book and in fact his later texts are still widely used (Jones, 1963).

Obviously, enormous advances in the subject have been made during the twentieth century. In thirties the U.S. Government provided large sums of money for studies and projects in the field of conservation, irrigation and flood control and Eagleson (1970) referred to this period as the 'First Golden Age of Hydrology'. Technological development during and immediately after World War II created new equipments and techniques, including improved versions of remote sensing equipments, and Eagleson felt this period as the 'Second Golden Age of Hydrology'. This new phase of research began in the sixties for finding rational solutions to the problems in water resources. At the present time hydrology is in a continual state of flux as new trends and new concepts develop (Ward, 1975).

Hydrological characteristics of river basins such as precipitation, runoff, groundwater table, utilization, development and management of water resources are studied by several scholars from geography and related disciplines. Considerable attention is paid to water resource evaluation techniques and methodology. The sources, extend and dependability of water supplies, the characteristics on which an evaluation of the possibilities for the utilization and control of water resources were studied by the UNESCO and WMD (1978) in its publication on Assessment of Water Resources. There are well known works carried out in this subject by Chow (1964), Kazman (1965), USDA (1969), Taylor (1970), Linsley and Frangini (1972), Ward (1975) and Uttan and Teclaff (ed.) (1975). A world survey on water resources is carried out by Furon (1967), Olson (1970) and Sewell (1975). There are many techniques of runoff prediction and forecasting. Some of these are in widespread use, either because they work reasonably well over a wide range of conditions or else are easy to apply. The use of other techniques may be restricted to specific areas or to specific users, such as a particular government agency.

Studies on drainage basin characteristics were contributed by Horton (1945), Strahler (1964 and 1971), Kulkarni et al., (1968), Daniel (1981) and Chang (1982). Gregory and Walling (1976) had studied the geomorphological approach of drainage basins. Contributions on hydro meteorological characteristics have been carried out by many. Natural evaporation from open water, bare soil and grass was studied by Penman (1948). Potential evapotranspiration was estimated by different scholars based on the modified book keeping procedure of Thornthwaite and Mather (1955 and 1957). Correlation analysis of hydrometeorological data was made by Sharane et al., (1976). Calibration for computing Thronthwaite's potential evapotranspiration in California was published by Shelton (1978). Hydro meteorological aspects of the Amazon basin had been studied by Hjelmfelt (1978). Other studies on hydrometeorological characteristics were made by Lvovich (1959) and Vendove (1960). Precipitation characteristics were dealt by El-Tom (1969), Haragan (1976), Zorne and Rogers (1978), Bowden (1980), and Johnson and Dart (1982). Precipitation over large areas was estimated by Thiessen (1911). The prospect of further developments in these fields will pose stimulating challenges to physical geographers and hydrologists during the next decade.

The ground water hydrology - concept and models and evaluation were studied by Baker and Horton (1936), Tolman (1937) Walton (1970), Todd (1971), Bouwer (1978) and Yamamoto (1979). Water table and its characteristics were studied by a number of scholars. Oskolkara (1979) had studied estimation of the changes of groundwater regime. Other related studies were Rarabough (1956), Akiyama (1980) and Suekane and Kayare (1980). Forecasting and mapping of groundwater table in USSR was studied by Zaltsberg (1982). Recharge estimation of groundwater were carried out by Lloyd et al., (1966) and Smith and Vikramaratna (1981). Conjunctive use of surface and groundwater resources were made by Brutsaert and Gebhard (1975) and Kos (1981). Application of statistical methods to the analysis of groundwater table was studied by Remson and Randolph (1958). New discoveries which continue to enhance the groundwater potential of the semi-arid areas not only suggest that irrigation will continue to play a vital role in the development but also underline the need for a better scientific understanding of irrigation problems.

Water balance analysis as developed by Thornthwaite and Mather offers a means of showing water deficit, water surplus and potential evapotranspiration over different areas (Thornthwaite and Mather, 1955 and 1957; Lvovich 1959; Yoshino,

1971; Yu, 1976; Peter, 1977; UNESCO, 1978 and Yoxall, 1980). Contributions on water resources development were made by United Nations ECAFE (1955 and 1961) and Ayoade (1975). Wiener (1971) had dealt with the role of water in development and demographic effect in water development was studied by Hollis and Envoy III (1971).

Ideally every public agency charged with implementing water management measures would like to be able to predict the effect of each contemplated action upon the interlocking system of land and water in the entire river basin and this requires and interdisciplinary approach. The availability, development and use of water resources had been studied and an approach to meet the future water requirements was outlined by Greet (1977) and Smith (1977). Optimum use of water resources was studied by Plera (1961) and De Ridder and Erez (1977). White (1972) had studied the role of geography in water resource management. This was directed towards the effect on the programme to attain regional economic development and placing first priority on enhancement of quality of the human environment.

Interpretation of surface water quality was dealt by Douglas (1972) and Walling and Webb (1975). Stream water quality as influenced by land use pattern was studied by Hirose and Kuramoto (1981). Studies on groundwater quality were contributed by Richards (1954), Nejand (1975) and Mostertman (1977).

Forecasting the demand of water for different purposes was studied by Sharp (1967), Rees (1973) and Batisse (1976) and Tate (1981). A new method of predicting long range water demands was developed by Maddam et al., (1979), to quantify the post draught effectiveness of water conservation in the San Francisco Bay area. Domestic water supply in the third world countries was studied by White (1978). Hydrological models and mapping in the context of forecasting have been made by Aston (1980). GIS for water resources and watershed management have been studied by John (2003).

The foregoing review reflects the sobering findings of investigations of what happened in the trail to water resources in earlier decades. In turn, it makes new demands upon scientific research. Heavier emphasis is placed upon research which cuts across the conventional disciplinary lines to find answers to newly phrased questions relating to environmental effects of human interventions in the hydrological cycle.

Water resources analysis helps to illuminate problems anticipating long-term fluctuations in water supply and choosing among a theoretically wide range of options for assessing environmental impacts. The medium and long range forecasts of stream flow should take into account the effects of change in watershed conditions as they may alter evaporation, transpiration, and soil storage and stream flow. In the long run, changes in agricultural and industrial land use may be of large consequences. Explicit provisions should be made in forecasting such changes.

The long term climatic and landscape changes need to be assessed more carefully. The tactics of manipulating water quality and quantity need to be appraised more fully. Special attention should be called to the possible consequences of major inter basin transfer of water. These problems should occupy a special place in future research on water resources.

Major improvement in future management could be gained by approaching water planning from a historical point of view. This means examining projected water management in conjunction with past modes of managing land, mineral and related resources. In each natural area it is possible to recognize distinct historical stages of economic development and resource management. Each had its own requirements for solving problems of water and environmental conditions.

1.3.1 Geographers' Contribution to Water Resource Studies

Geographers have long regarded the study of water resources as an essential part of their discipline. Also there has been shift in their interest on the studies of water and Ward (1978) drew attention to parallel shift of emphasis in a number of countries from a dominantly physical concern with the natural water cycle to an increasing awareness of and concern with the socio-economic implication of hydrology.

Regional studies and studies on settlements make reference to water and a large number of research papers describe systems of water use for both domestic and irrigation. A volume of essays in the field of hydrology, geomorphology and socio-economic geography give the views of geographers on the inter relation of hydrologic, geomorphic and human systems and notes some contributions geographers can make to actual water management (Chorley and Kates, 1969). The problems of water resources were dealt by Hidore (1970). Contribution of geographical analysis to river basin development was made by White (1963). An international experiment in water resources development in the Lower Mekong basin was unique in seeking to relate the full array of scientific knowledge and method of planning (White, 1963). The other related studies on the Lower Mekong basin were made by Johnson (1969), White (1969) and Ganesh (1984).

Soviet Geographers had carried out several detailed water resource appraisals, and a number of them have been active in the hydro meteorological institute under Davitiya (Vendrov and Kalinin, 1960; and Vfremove, 1961). Soviet Geographers, with regard to resource estimates, helped improvement in mapping of water balance in the Soviet Union (Lvovich, 1959). There are studies giving a computation of the amount of water available and required for plant growth in dry areás. Such estimates are likely to be less exact than those based upon adequate stream flow from short term and sporadic observation.

The most systematic work towards the development of science of groundwater hydrology was carried out by a French investigator Belgrand, who made the distinction between permeable and impermeable formations. Darcy established the law of flow of groundwater. Dawbree made a significant contribution to the subject of the relation of geological structure to the occurrence and movement of groundwater (Mathur, 1969).

Polish geographers have presented hydrographic mapping of both surface and groundwaters assiduously and are moving ahead with comprehensive mapping of these characteristics for the entire nation (Klimaszewski, 1956 and Leszezyeki, 1960).

Japanese scholars are also quite active in the field of hydrology. They established a commission on hydrology in 1967. Yamamoto (1986) had discussed the relation between ground and surface water using data of groundwater table and river discharge of the Shonai River. There have been studies on water balance too (Keller, 1968; Yamomoto, 1970; Yoshino, 1971). Akiyama 1981 made a study on recent changes of water uses.

Geographical analysis has played an important role in unfolding a set of approaches to river development and at present joins the few systematic efforts to examine what has happened as a result of investment in river development and what lessons may be learned for future ventures in the field. Contributions on water resource appraisals by geographers are modest. The geographical skill has been to combine knowledge of physical parameters of the water resources with insight into ways in which data can promote wiser choice among possible adjustment to possible time pattern in resource availability. Resource estimated can be improved by refining the mapping of water balance, hydrological characteristics of terrain and land use and by examining the nature and distribution of human perception of resources. The application of technology can be strengthened by discovery and partial explanation of variations in patterns of water use and disposal. Benefit-cost analysis can be sharpened by finding normalities in the spatial variations of water use, control and demand. The effects of water management upon the stream systems and upon movements of human activities within and outside the basin can be identified more clearly (White, 1972).

There are three major problems which emerge from geographic analysis of water resources, namely 1. Assessment of the possibility and implication of long-term fluctuations in the supply of water, 2. The narrow range of possible adjustments which is commonly canvassed in planning the changes in spatial patterns of water, and 3. Predictions of environmental consequences due to water development. Each of these has implications for organization of the scientific and administrative activities relating to them.

1.3.2 Water Resource Studies in India

The geography of water resources in relatively a new subject in India. The most important schemes relating to water resources development were conceived and executed after Independence. It is not surprising, therefore, that most of the geographical literature existing on the subject and most of the publications relate to the period after 1950. Indian geographers have been studying the problems of water resources in the country with a view to find out ways and means of scientific utilization and conservation of water.

Chatterjee (1968) had reviewed in brief the status of hydrological studies in India till the mid-sixties especially with reference to floods and flood controls, study of lakes, springs and rivers and irrigation and groundwater studies. Shafi (1972) had examined the status of Indian studies concerning quantitative and regional surveys and mapping of utilization of water resources, status of tube well irrigation and the problems of flood and flood control. Dayal in the same survey reviewed the progress of studies in river valley development planning in the country while Chatterjee and Lahiri reviewed contributions to the knowledge of natural hazards like floods and draughts in India, water pollution and problems arising out of injudicious use of irrigation water. Arunachalam (1983) made a detailed survey of water resource studies in India.

Significant studies have been carried out in the field of hydrological characteristics of river basins such as precipitation, runoff, availability, utilization and demand of water resources. Srinivasa Rao et al., (1971) had studied the various techniques and methodology for assessing surface water resources of a basin. Evaluation of water resources were made by Khosla (1951), Singh (1959), Rao et al., (1971), Kumaraswamy (1974), Sen (1976) and Subrahmanyam and Viswanathan (1979). Sharma (1976) had studied some hydrological characteristics of the Damodar River. Singh (2002) had analyzed hydrological Challenges of India. Trivedi et al (2005) had dealt the hydrological studies of Ghuradongri Watershed of Betul District, Madhya Pradesh.

Estimating potential evapotranspiration by different methods had been studied by Padmanabhamurthy and Biswas (1970) and Subramanian and Rao (1980). The hydrometeorological characteristics of Narmada basin were studied by Abbi et al., in 1970. Sub-basin wise run off had been estimated by Narayan (1976) and Reddy and Reddy (1984). Floods and its impact on regional economy were studied by Bannerjee (1944), Sirkar (1954), Chakravorthy (1959), Chatterjee and Bagchi (1961) and Goswamy (1972).

Groundwater studies have been receiving much greater attention in India, of late, both from State agencies and individual workers. Sharma et al., (1971) had studied the various techniques and methods for development and assessment of groundwater resources of a basin. Groundwater resources were evaluated based on climatic appraisals by Subrahmanyam and Sastry (1971) and Subrahmanyam et al., (1979). Bilas (1980) had studied the groundwater resources of Varanashi district. Availability of groundwater in various states were studied by Mathur (1969) Sharma (1978) and Rai and Dutta (1982). Groundwater table and its characteristics were studied by Gupta (1976) and Singh (1978). Chaturvedi (1977) had examined the planning for groundwater resources. Exploration and exploitation of groundwater with reference to recharge was studied by Pathak (1978). Datta et al., (1979) had dealt with the comparative study of groundwater recharge rates in parts of the Indo-Gangetic and the Sabarmati alluvial plains. Reddy et al (2000) had studied evaluation of groundwater potential zones using remote sensing data.

Using the modified water balance procedure of Thornthwaite and Mather, yield from river basins had been estimated by Rao and Subrahmanyam (1968) and Subrahmanyam (1982). Studies of water balance for river basins were made by Subrahmanyam et al., (1970 and 1980) for the Godavari and the Krishna basins. Water balance and its applications in different areas were carried out by and Subramaniam and Rao (1979). The study of droughts based on water budget method seems to provide a more realistic approach than that based on rainfall alone. This had been studied by Rakhecha (1974) for Andhrapradesh. Contributions of water balance studies for different areas of India were dealt were by various scholars (Rathor and Singh 1970; Sharma, 1974; Rao et al., 1976; Dey, 1978; Subramaniam and Uma Devi, 1976; Subramaniam and Rao, 1979; Subramaniam and Rao, 1980; Subramaniam and Vinayak, 1982 Madhuramma (1992) and Hemamalini (1992). Ramanaiah and Ravindrareddy (1997) had analyzed surface water balance of Chitravathi River.

A few significant studies have been carried out in the field of water resources development in India. Bose (1948) dealt with all aspects of the Damodar Valley Authority Project and Kirk (1950) presented a comprehensive and balanced regional geography of the Damodar basin. Other related studies for different areas were made by Ray (1950), Hart (1956), Gokhale (1959), Venkateswaralu (1959), Sharan (1960), Naegamvala (1961) and Jain (1968). Hydel Power generations schemes of various Indian rive have been studied by different scholars. While examining the hydrology of the west flowing rives Shankarappa et al., (1968) had analyzed various hydroelectric schemes of these rivers of Karnataka state. The other studies were made by Krishnamurthy (1940), Dutt (1948), Jauhani (1956) and Ramamoorthy (1963). Water for India in 2050 had been analyzed by Gupta and Deshpande (2004).

Water resources utilization and its problems were studied by Saxena (1974), Rao (1975), Bohle (1976) and Chaturvedi (1976). Singh (1959) dealt with available water resources and demands for irrigation, flood control, power generation and the need for domestic and industrial uses. Contributions with regard to irrigation were made by many. Theory and practice of irrigation was dealt by Michael (1978). Singh had examined the irrigation water and its impact on agricultural production in reservoir command areas were discussed by Sharma and Jagdish Kumar (1979). A historical account on irrigation and water supply in South India up to 1300 A.D. had been analyzed by Srinivasan (1968). Abhu and Salvia (1993) analyzed the utilization of water resources for agriculture in western Uttar Pradesh. Bhati (1996) had analyzed water resources and their utilization in India.

Water resources management for different command areas and states were studied by Ali (1980), Mathur and Yadav (1980), Rao and Kumar (1980) Sivanappan (1980) and Basil (2004). . The distribution of water resources and water management in the drought prone areas of Maharastra were examined by Gandhi (1979). The probability of water surplus and deficiency had been utilized in viewing the problems of agricultural water management. Patel (2002) had studied water resources management and people's participation. Chandrasekharan et al (2004) had analyzed the water resources development and management in India. GIS based technologies for watershed management studied by Gosain and Rao (2004).

The quality of water from the point of view of the injurious of irrigation and post-irrigation had been studied by many. (Biswa, 1979; Sarma and Swamy,1981; Sudhakar and Mamatha, 2004; and Madhuri and Geetika (2005). Basin-wise groundwater quality was studied by Niyogi and Bagchi (1981) and Sivagnanam and Kumaraswamy (1983). The effect of water table fluctuations on water quality had been studied by Rao (1978). Rural drinking water supply was studied by Balakrishna and Reddy (1978) and Bilas (1981). The development of inter-state rivers was studied by Sain (1977) and water disputes between states were studied by Sukwall (1979). Using aerial photographs and satellite images hydrological investigations were made by Sharma & Jain (1978), (1981) & Thiruvengadachari (1981) Srivatsava , Cahrya and Amit (2000).

Studies on water resources with respect to Tamil Nadu are not may. Water resources of Tamil Nadu were studied by Rao in 1968. A water atlas had been prepared by the Government of Tamil Nadu (1974). Hydrological characteristics of different basins were investigated by Ayyappaswamy (1978), Chinnamani and Sakthivadivel (1981). Kumaraswamy (1978) had developed a mathematical model for river basins of Tamil Nadu. This is useful for optimum design of water resources and also for the optimum design of water resources and also for the efficient management of existing water resources. Surface water availability was also examined by him (1974). Investigations of groundwater availability were dealt by Balakrishna et al. (1972), Panchanathan (1976), Public Works Department (1976) and Central Ground Water Board (1980 and 1983). With regard to water balance Anandaraj (1978) and Ram Mohan (1978) made detailed studies on Tambraparni and Tamil Nadu as a whole respectively. The geographical analysis on the Palar basins was made by Thirunaranan (1938). The development of irrigation was made by Anandarao (1958). Conservation of surface water and modernization of water resources development schemes were studied by Vedanayagam (1977). Jeyapalan (1979) had dealt with water management and tank irrigation systems in

Tamil Nadu. Water shed management in Chinnar reservoir catchments was made by Kalpalatha (1983). Sankar made a study on evaluation of groundwater potential zones using remote sensing data in Upper Vaigai basin (2002).

On the whole, there has been a concerted effort by Geographers. They have advanced a little in their methods of estimation of availability and utilization of water resources. Only recently, work in theoretical geography has influenced the conceptual approach to water problems. The development in this area of study should cover three major aspects: 1. more theoretical approach and less concern with the uniqueness of geographic distribution; 2. a change from the deterministic interpretation of phenomena to probabilistic interpretation; and 3. the micro-geographic study.

However, it is surprising to see that in Tamil Nadu the aspects of resource appraisal have not drawn much attention. Owing to the population pressure and irrigation development it is imperative to study the water resources of a region. Tamil Nadu has already utilized more than 95 per cent of the existing surface water resources (Kumaraswamy, 1974). With the increase in population and rising demands for water for irrigation, industrial and other purposes, water is in fact a scarce commodity in the State and thus the scientific development of water resources has assumed growing importance in the economic and social development of Tamil Nadu. Considering the cardinal importance of water resources it is surprising that this aspect of study has not received enough attention from the scholars in Tamil Nadu.

1.4 UPPER VAIGAI BASIN: STUDY AREA

Tamil Nadu, one of the States of the Indian Union, is located in the southeastern part of the Indian Peninsula. The areas of high rainfall are located along the Western Ghats of Tamil Nadu running from north to south. Rainfall is scanty on the leeward side of the Western Ghats comprising mostly the interior parts of Tamil Nadu. The broad stretch of central Tamil Nadu receives rainfall between 750 to 1000 mm, which could be considered as low and insufficient for agricultural proposes in the prevailing agro economic conditions.

The State receives the bulk of its rainfall during the northeast monsoon season. As a matter of fact, much of the agricultural rhythm in Tamil Nadu coincides with this season.

Rainfall in Tamil Nadu is highly variable and unreliable. Consequently droughts and famines are very frequent in the State, particularly in the southeastern parts. The State has a very heavy demand of moisture due to high temperature all around the year. The distribution of water surplus indicated that the vast stretch of the semi-arid interior has now water surplus at all. The hilly region on the west records a surplus between 600 and 800 mm whereas in the sub-humid zone along the north Tamil Nadu Coast a small water surplus of about 200 mm occurs.

The surface flow consists of direct runoff from rainfall and flow in streams and water regenerated from groundwater flow in Tamil Nadu is 4.1 million ha. m. It includes 3.25 million ha. m from rainfall 0.75 million ha. m transferred from river basins of Karnataka and Kerala and 01.10 million ha. m from irrigated fields. Of the total surface water of 4.1 million ha. m about 1.5 million ha. m is stored in reservoirs and tanks. Of the 2.6 million ha. m that flows in the rivers and streams utilization of water works out to 2.45 million ha. m. The balance 0.15 million ha. m flows to the sea as flood flow (Institute for Water Studies, 1984).

Contribution of groundwater from rainfall is 1.4 million ha. m. With the addition of 1 million ha. m from irrigation system and 0.15 million ha. m from the streams the total groundwater potential is 2.55 million ha. m. Out of this potential 1.15 million ha. m is extracted annually and about 0.55 million ha. m is extracted annually and about 0.55 million ha. m is regenerated into rivers and the balance is accounted for by increased groundwater storage and evaporation (Institute for Water Studies, 1984).

The State is dominated by the Cauvery System, which drains more than one-third of the area of the State. The other important basins are the Pennaiyaru, the Palaru, the Vellaru, the Vaigai and the Tambraparni.

The Upper Vaigai basin is situated in the Theni and Dindigul districts. Covering an area of about 3975 km^2, it had a total population of 10, 89,785 in 1981. The basin is located in between the Lower Palanis, the Cardamom hills and the Varushanad-Andipatti hills of the Western Ghats. The basin is drained by a few main tributaries of the Vaigai namely Suruliar, Kottagudi, Varahanzadi, Manjalar and Maruthanadi.

Population distribution is quite uneven and is concentrated in areas along the Suruliar, the Kottagudi, the Varahanadi, and the Manjalar. It is sparsely distributed in the sub-basins of Thevaram, the Andipatti the Viruveedu and the Thalai Vaigai. About one third of the population lives in urban areas.

An uneven distribution of settlements is also noticed. The Varushanad-Andipatti hills, the Lower Palanis and the Cardamom hills have only a few settlements. All along the river courses of the Suruliar, the Kottagudi, the Varahanadi and the Manjalar sub-basins the concentration of settlements are very high. About 38 per cent of population constitutes the workforce of this basin. Among this three-fourth are engaged in primary activities. Population engaged in secondary and tertiary activities are very meager in the rural areas in contrast to the urban areas where it is a high as 36.9 per cent.

The Upper Vaigai basin lies on the lee side of the Cardamom hills and so it receives less rainfall (mean annual 883 mm). The highlands of the Lower Palanis and the extreme southeastern part of the Varushanad hills receive the maximum rainfall in the basin (1200 mm). The central part of the basin proper running southwest to northeast is a dry tract receiving less than 700 mm of rainfall per annum. Except the dry central stretch the rest of the plain areas near the foot hills receive rainfall ranging from 700 to 800 mm per annum. The southwest and northeast monsoon seasons account for nearly 80 per cent of the rainfall. Between the two monsoons the northeast monsoon is more important accounting for nearly 45 per cent of the annual rainfall while the southwest monsoon contributes about 35 per cent of the rainfall in the basin. A very high coefficient of rainfall variability and thus low rainfall stability is experienced in the western part covering the Thevaram sub-basin. A very low coefficient of variability and greater rainfall dependability are noticed in the Lower Palanis and the Varushanad hills.

The Upper Vaigai basin, as in other parts of Tamil Nadu, has developed its surface water resources to the maximum extent. A substantial portion of the ground water resources has also been utilized. In such a situation future demands are to be met by judicious management of water resources. To formulate a plan for efficient management of water resources it is necessary to make an assessment of their quantity, distribution and utilization.

1.5 THE RESEARCH PROBLEM

The Upper Vaigai basin is one of the river basins of Tamil Nadu located in a rain-shadow region of Theni and Dindigul districts. With less than 883 mm of rainfall annually the basin faces a critical situation in respect to water-use. Population is heavily concentrated in the central part of the basin and as such a water need is the maximum in this belt. Consequently this region faces acute water deficiency which is as high as 900 mm PE > 1600 mm and AE < 700.

Studies available on the Upper Vaigai basin are very limited. Studies on the Pereira Vaigai Irrigation projects were dealt by World Bank (1977) and Vedanayagam (1968). Rashid Ahmed (1976 and 1977) had examined the power projects in Madurai district. National Remote Sensing Survey (1979) has completed a Satellite Remote Sensing Survey of this basin during their survey of the southern part of Tamil Nadu. However, comprehensive studies on the spatial and temporal distribution of water resources, its availability, utilization and development are yet to be made.

The availability, utilization and development of water resources are not same in all the sub-basins due to variations in their size and location. Due to these differences there seems to be a striking disparity in the economic development between these sub-basins. A clear understanding of the water availability, utilization and problems and prospects of water resources of each sub-basin will be very helpful for planning and development of water resources which in turn will be quite helpful for the betterment of the society living in the basin. The present geographical research goes for this kind of study.

1.6 AIMS AND OBJECTIVES OF THE STUDY

The study aims to appraise the availability, utilization and development, and problems and prospects of water resources in the Upper Vaigai basin Tamil Nadu. The major objectives are:

1. to evaluate the surface and groundwater availability of the Upper Vaigai basin, taking up each sub-basin,
2. to estimate the water balance of the basin and to identify the water deficit and surplus areas,
3. to study the patterns of water resources utilization and development in the study area and
4. to examine the problems and prospects of water resources in the basin.

1.7 METHODOLOGY

1.7.1 Date Base

The study is based on the information from primary and secondary sources. The climatic data such as rainfall, temperature, relative humidity, wind speed and sunshine were collected from the India Meteorological Department and Tamil Nadu Government Research Stations; hydrologic data such as stream flow, groundwater level, specific yield and data on groundwater exploration were collected from the Gauging and Groundwater Divisions of the Public Works Department; Irrigation data regarding dams, anicuts, canals, tanks and wells and ayacut were collected from the Irrigation Division. The village wise data on land use and irrigation were collected from the Revenue and Statistical Departments. Particulars regarding urban and rural water supply were collected from the Tamil Nadu Water Supply and Drainage Board. Data were also generated from the recently published Survey of India topographic sheets, geological and soil maps and landsat and IRS 1C images. Primary data were collected from sample villages in the study area in order to study the problems associated with water resources.

1.7.2 Analysis

A number of techniques of water resources evaluation have been adopted to study the water resources of the basin. Employing the book-keeping procedure of Thornthwaite and Mather, water balance is estimated and areas of water deficiency and surplus are identified. The stream flow data are assessed using log-normal and Gumbel distributions for the annual flows, and maximum, minimum, mean, standard deviations and flow duration curves for the monthly flows. The groundwater yield is estimated through specific yield and rainfall recharge methods. Trend surface analysis is used to study the overall trend in mean annual rainfall distribution and groundwater table. Appropriate cartographic techniques have also been applied wherever necessary.

1.8 ORGANIZATION OF THE STUDY

The present study on 'Water Resources Evaluation Methods and Techniques', is organized in five chapters. The first chapter unfolds the subject matter and procedure of the study. The second chapter gives a geographical background of the Upper Vaigai basin. Evaluation of surface and groundwater resources along with climatic water balance is analyzed in the third chapter. The fourth chapter examines the development and utilization of surface and ground water resources and provides an insight into the problems and prospects of water resources. The final chapter summarizes and outlines of the future course of study.

CHAPTER 2

UPPER VAIGAI BASIN: A GEOGRAPHICAL BACKGROUND

2.1 LOCATION

The Upper Vaigai basin spreads over an area of 3976.35 km^2 in the Theni and Dindigul districts of Tamil Nadu. It lies between 9° 31′N to 10° 21′ N latitudes and 77° 10′ E to 77° 51′ 15 E longitudes. The basin covers Uthamapalayam and Periyakulam Taluks in Theni district and parts of the taluks of Kodaikanal, Dindigul and Nilakottai in Dindigul district (Fig. 2.1).

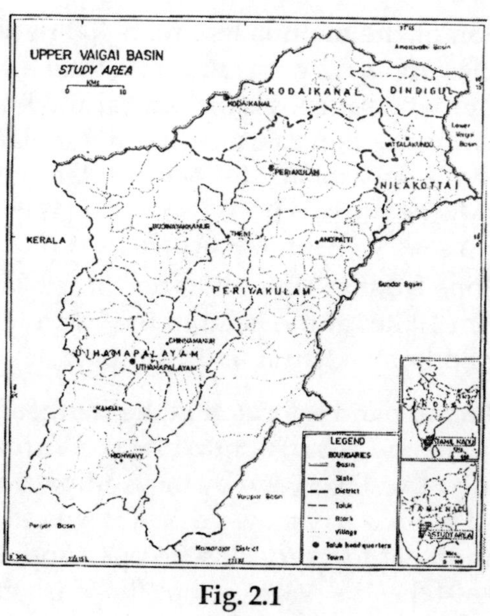

Fig. 2.1

The basin is surrounded to the north by the Amaravathi basin; on the northeast by Cholavandan sub-basin of the Lower Vaigai basin; on the east and southeast by the Gundar and the Vaippar basins; and on the west and southwest by the Periyar basin.

The study area consists of 129 villages and eleven towns (Bodinayakkanur, Theni-Allinagaram, Kambam, Periyakulam, Gudalur, Chinnamanur, Kodaikanal, Uthamapalayam, Vattlagundu, Andipatti and Highwavys). The total population in the basin was 10, 89,785 persons according to the 2001 Census with an overall density of 274 persons per square kilometer.

2.2 PHYSICAL BACKGROUND

2.2.1 Physiographic Divisions

The altitude varies from 200 m in the northeastern part of the basin to 2640 m in the northwestern part of the basin (Fig. 2.2). Within the basin, three broad physiographic regions may be identified: the plains in the centre, the plateau bordering the plains in the centre, the plateau bordering the plains in the southwest and northeast, and the hills forming a rim of highlands all around the basin.

2.2.1.1 The Plains

The elevation of the plain is less than 400 m above mean sea level and occupies about 42 per cent of the total area of the basin. The 400 m contour which encloses the central and northeastern part and the basin forms the boundary between the plain and plateau. Here and there are a few isolated hillocks. Notable among them are the ones near Markayankottai (700 m), Nagalapuram (706 m), Kulipatti (713 m) and Senkattampatti (701 m) (Fig.2.2). The plain has a general slope from southwest to northeast. The width of the plain ranges from 10 km at the head of the basin to 50 km along a line connecting Bodinayakkanur and Andipatti.

The Varushanad plain is located in the southeastern part of the Upper Vaigai basin and is surrounded by the Varushanad hills on the south-southwest and west and by the Andipatti hills on the east. The Varushanad plain is a narrow neck of low land due to the hills on either side approaching very close to one another rat this point. 'The river Vaigai takes its source from the Varushanad hills and descends the slopes of these hills and enters the Varushanad plain and flows north-northeast from Kovilankulam' (Muthuswamy, 1932).

To the west of the Varushanad plain is a rich fertile plain, the Kambam plain, drained by the river Suruli and its tributaries. It occupies nearly 35 per cent of the study area. To the west and northwest are the Thevaram and Theni plains drained by the Thevarm and the Kottagudi rivers respectively (Fig.2.2). The general slope of these plains is towards the east-southeast and this is the region dominated by both tank and well irrigation. The other sub-basins Varahanadi, Manjalar and Maruthanadi are located in the north of the study area and the general slope of these plains is from northwest to southeast. The north and northeastern part of the Upper Vaigai basin has more number of canal-fed and rain-fed irrigation tanks.

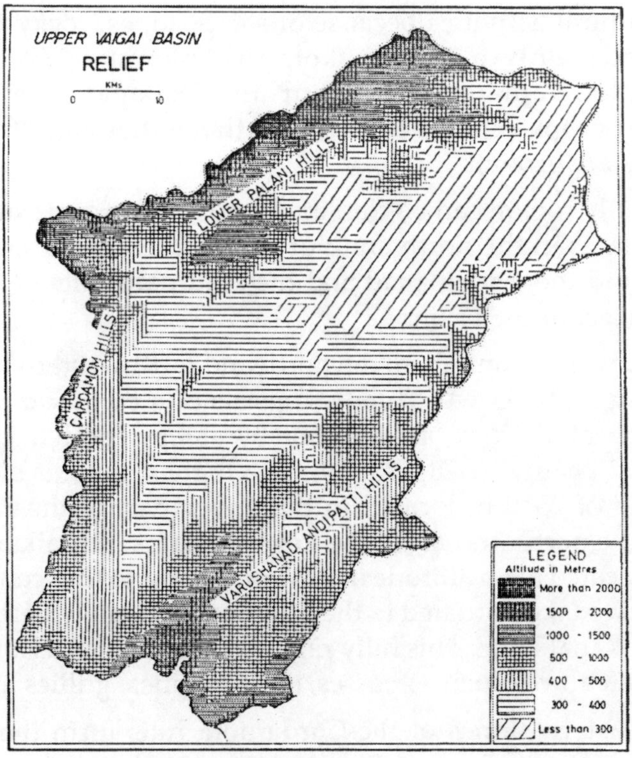

Fig. 2.2

2.2.1.2 The Plateau

Extending continuously from the Upper Suruli basin to the Upper Thevaram basin and the Varushanad Valley, is located the plateau region with an average width of 40 km and an elevation of 400-500 m above mean sea level. This forms the transitional zone between the plains drained of the Upper Vaigai system and the forest clad hills with steep slopes above. The plateau coves nearly 18 per cent of the total area of the basin. Though the local relief is a plateau, it does not completely hinder agricultural activities as the plateau is quite broad.

2.2.1.3 The Hills

The area above 500 m is classified as the hilly region. It covers nearly 40 per cent of the total area of the Upper Vaigai basin and forms a rim all round. The hilly region is marked with numerous cliffs, ravines, gullies and escarpments. The whole area is unsuitable for agricultural activities because of steep slopes, rocky outcrops, ravines etc. Only a few pockets in Agamalai, Kodaikanal, Pannaikadu, Kamanur, Poolathur and Vellagavi are used for plantation crops while near around the settlements fruits and vegetables are grown.

The hilly region can be classified into three groups such as the Lower Palanis in the north and northwest; the Cardamom hills in the west and the Varushanad and the Andipatti hills in the south and southeast of the basin.

The Lower Palanis form the north and northwestern boundary of the basin. 'The Lower Palanis consists of a confused jumble of peaks separated from one another by steep wooded valleys of exceptional beauty' (Baliga, 1960). The highest peak is situated at an altitude of 2519 m located in the upper part of the Kottagudi basin. In the northern part of the Lower Palains, Kodaikanal, a hill resort, is located at an altitude of 2234 m. To add to the scenic beauty there are two lakes situated in the hilly region namely Berijam Lake and Kodaikanal Lake. This hilly region consists of an assortment of landscape features such as passes, cliffs, ravines, gullies and cols.

The eastern slopes of the Cardamom hills form the western boundary of the study area. The slope is indeed very much steeper on the eastern side. 'The drainage pattern on either sides present some very remarkable adjustment to structure, and this is an equally remarkable discordance' (Spate, et al., 1972). The highest peak in the Cardamom hills within the basin boundary is 1589 m above mean sea level near Silamalai.

The Varushanad and the Andipatti hills are located in the south and southeastern part of the basin. The maximum altitude is 1896 m in the southern most tip. Another high peak is seen at Kovermalai at an altitude of 1754 m. The western portion of the Varushanad hills is known as 'Highways'. 'The higher altitudes of these mountains are covered with thick forests and further down a line of barren land and rough grass can be seen while at the foot hills extensive forests abound' (Baliga, 1960). The maximum altitude in the Andipatti hills is 1301 m. The Andipatti hills are comparatively lower and have gentler slopes. They are characterized by a few spurs too (Pemalai, 1700 m; Suddragiri, 1272 m; Kudiramalai, 1700 m). In places, they are more broken and scattered as 'blocks of naked granite' (Muthuswamy, 1932).

2.2.2 Geology

The different geological formations of the Upper Vaigai basin are broadly classified into crystalline rocks of the Achaean era and sedimentary formations of the quaternary era (Murthy, 1971 and 1975). Charnockite garnetiferous sillimanite gneiss and biotite gneiss are the predominant rock types of the archaean complex enclosing bands of calc gneiss and quartzite and intruded by granite, pegmatite and quartz vein, whereas calcareous mud and clay, aeolian sands and soils belong to the quaternary era.

The geological succession of the rock formations of the basin is presented in Table 2.1.

2.2.2.1 Calc Gneiss

Calc gneiss outcrops are seen in the biotite gneiss, and charnockite mainly on the eastern part of the Kambam Valley (Fig. 2.3). A small outcrop occurs also near Thimmarasanaickanur. The band of the calc gneiss is in conformity with that of the enclosing rocks. The width of the calc gneiss bands varies from 10 to 30 m and their length from 200 m to 10 km. The bands are located from 200 m to 10 km. The bands are located near Andipatti, Kamayagoundanpatti, Narayanathevanpatti, Surulipatti, Karunakaramuttanpatti, Pandarathurai and Kanavoypatti.

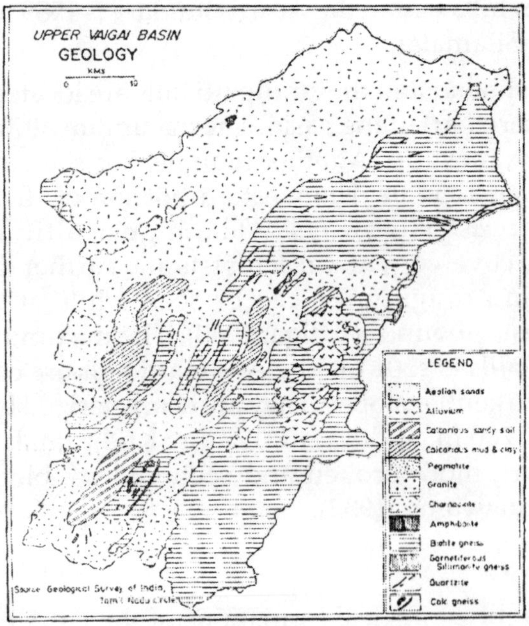

Fig. 2.3

Table 2. 1 Upper Vaigai Basin: Geological Succession

Quaternary	Soils
	Aeolian sands
	Calcareous sandy soils
	Calcareous mud and clay
	Pegmatite and quartz vein
Archaean	Granite
	Charnockite
	Composite gneiss
	Amphibolites
	Biotite gneiss
	Garnetiferous sillimanite
	Gneiss
	Quartzite
	Calc gneiss

2.2.2.2 Quartzite

A number of parallel, linear bands of quartzite varying in width from 10 to 100 m and in length from 0.4 km to 5.0 km are found in the plain areas of Uthamapalayam and Periyakulam taluks. Quartzite is also noticed near Amatchiyapuram, Marikundu, and Pichchampatti. They generally stand out as erosional ridges without much vegetation except for thorny shrubs (Fig. 2.3).

2.2.2.3 Composite Gneiss

Composite gneiss consisting of amphibolites, biotite gneiss and garnetiferous sillimanite gneiss occur as large bands and are next to charnockite in the areal extent. Amphibolites are noticed as thin bands and lenses, within the charnockite rocks near Kodaikanal. They probably represent the xenoliths of the basic flows and sills into the original Archaean sediments, which have suffered repeated metamorphism, folding etc. The amphibolites are 0.3 m to 2 m wide and extend for 5 m to 8 m in length. Predominant occurrence is noticed on the north of Kodaikanal.

Biotite gneiss in the more predominant type and the presence of grant and sillimanite appear to be a local variation, as seen on the southwestern slope of the Highwavy mountains. These gneissic rocks are well exposed in the hills and the plain country east of Chinnamanur and west of Uthamapalayam (Fig.2.3).

2.2.2.4 Charnockite

Charnockite forms the major rock type of the basin. It is formed on the northwestern and western part of the basin (Fig.2.3). The plain country to the south and southwest of Venkatachalapuram and north of Pudupatti is mainly occupied by charnockite with associated parallel runs of meta-sediments like composite gneiss and quartzite. Charnockite also occupies the plain country to the west of Andipatti and also occur as ventricular bodies and bands in the composite gneiss which is the predominant rock type in the area. Also bands of charnockite are seen folded along with composite gneiss and calc gneiss to the east of Thimmarasanayakkanu where it occurs on the outer flanks of the plunging syncline. In the area around Andipatti the charnockite is seen only in the cross sections in wells.

2.2.2.5 Granite

Charnockite and composite gneiss are intruded by granite. The important localities are seen near Kamayagoundanpatti, Narayanathevanpatti, Surulipatti and Karunakaramuttanpatti. In general granite is pink in color and is medium to coarse grained. It is composed of quartz, pink feldspar, biotite and hornblendes.

2.2.2.6 Pegmatite and Quartz Vein

Pegmatite of granitic composition traverse the composite gneiss in charnockite. The important occurrences are seen north of Kumili. The width of the veins run from 20 m to 50 m and in length from 200 m to 500 m. Quartz vein occurs in the composite gneiss and in the charnockite near Mangaladevi hills. In general they are very thin.

2.2.2.7 Calcareous Mud and Clay

After a long period of quiescence the area experienced sedimentation, represented by the fresh water calcareous sediments amidst a group of crystalline rocks. These sediments have been seen west of Pulikuthi and east of Veppampatti (Fig. 2.3). Probably they are the sediments brought out from the adjoining hills and deposited in a basin like depression. The average width of the formation is about four km. The trend of the basin is northeast - southwest which happens to be the regional trend of the crystalline rocks.

The formation east of Veppampatti is one kilometer in average width and surrounded by a number of discontinuous ridges on the north and west and by the Varushanad hills on the south and east. In general the quaternary formation consists of horizontally bedded calcareous mud and clay. The rocks are generally white to grayish white in colour.

Calcareous mud and clay forms a fresh water sequence of sediments and is lying amidst the composite gneiss and charnockite in the Thevaram sub-basin. The fresh water sediments occupy an elongated portion in the northeast - southwest direction whose length is about 20 km and width is about 6 m to 8 km. The average thickness of the sediments varies from 2.5 m to 5 m.

2.2.2.8 Calcareous Sandy Soils

Calcareous sandy soil overlies the group of crystalline rocks of the Archaean age mainly in the Kamban Valley. They may represent a sequence of transported sediments. As the calcareous sandy clay is encountered only in the dug well section, no sedimentary features could be observed. They are seen in the south of Kambam to the west of Karunakaramuttanpatti and to the east of Gudalur. The maximum width of this formation measured in the east west direction is 6.3 km.

2.2.2.9 Aeolian Sands

Aeolian sands are noticed around Thevaram, Pottipuram, Silamarathupatti and Silamalai. They consist of yellow or brown and buff coloured, very fine grained, fine sandy clay to dust particles and extending to an area of 6 to 8 km. Occasional migration of san dunes is evidenced by their encroachment on productive fields.

2.2.2.10 Soils

Brownish black clayey soil overlies the charnockite and gneissic rocks of the basin. Reddish brown sandy soil admixed with rock debris is seen in the quartzite ridges.

2.2.2.11 Structural Features

Variations in the strike and dip of foliation of the various Archaean formations are seen in the basin which may be due to the intensive and repeated metamorphism. In some places they are tightly folded and crumpled. Near Periyakulam and Uthamapalayam the regional strike conforms northeast - southwest with moderate to steep dips on either sides.

Evidences of lineaments faulting are available from photo geological interpretation and field observation in many places. Indication of a regional fault plane running northeast - southwest direction is shown in Fig.2.4. In the charnockite evidences of faulting is observed near Vanathiparai around Mangaladevi hills. Joints and fractures are found in the rock types such as quartzite, gneiss, charnockite and granite. Strikes and joints are also predominant in garnetiferous gneiss, biotite gneiss and occasionally granites.

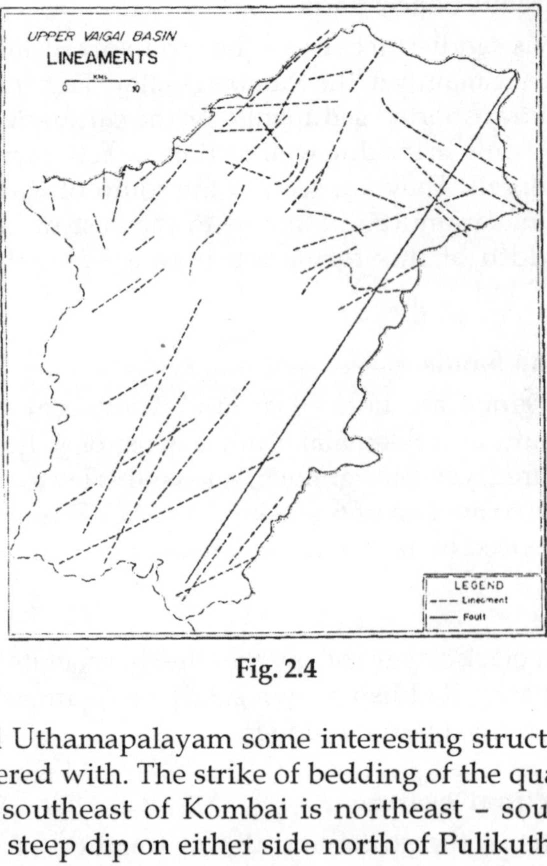

Fig. 2.4

Around Uthamapalayam some interesting structural features are encountered with. The strike of bedding of the quartzite bands in the area southeast of Kombai is northeast - southwest with moderate to steep dip on either side north of Pulikuthi conversing dips in the quartzite band indicate a syncline trend to the northeast. This syncline section extends up to Kulayanur. As such the syncline extends from 2.4 km to the northeast up to a hill 765 m high and to 2.1 km north of Kulayanur.

An anticline could be traced to the east of the syncline described above. The direction of the axial plans of the anticline is northeast - southwest. Another syncline with its axial plains running in a north northeast - south southwest direction is noticed about 1.8 km east of Rayappanpatti. This fold continues up to Chinnamanur. About 2 km to south of Chinnamanur, a folded calc gneiss band is noticed.

The strike of the band is northeast - southwest in the western limb and north northwest - south southeast in the eastern limb. The dip of the band varies from 50° to 60° on either side. Another syncline is identified based on the altitudes of the band. This is followed towards the east by an anticline. About 9 km to southeast of Kamayagoundapatti an anticline fold is deciphered based on the attitudes on quartzite bands. A synclinal fold with its axial plane extending north-south is delineated 2.2 km south-east of Andipatti. Another syncline fold is seen about 0.3 km east of Thimmarasanayakkanur with its axial plane in a north-south junction.

2.2.3 Climate

Most of the basin has a semi-arid type of climate. The north and northwestern part of the basin fall under per-humid type of climate due to the high altitudes based on Thornthwaite's scheme of classification.

This basin shows a large variation in temperature because of the variations in altitude. From Appendix 2.1 it is observed that the summer season has recorded a very high temperature. The mean annual temperature of Kodaikanal, Kamatchipuram, Viralipatti and Madurai are 14.2, 26.2, 28.4 and 28.6°C respectively. Since Kodaikanal is located at an altitude of 2248 m the temperature never exceeds 16.16°C.

The weather of the basin becomes drier as one proceeds from south to north. The southern part especially Melagudalur, Kambam and Uthamapalayam enjoys a typical monsoon climate, while the lands north of Veerapandi do not enjoy. The summer heat is reduced by the influence of the southwest monsoon. When the monsoon breaks there is a heavy drizzling throughout the Kamban Valley from the Periyar Lake to Veerapandi (Ayyengar, 1933). The weather becomes drier in the northeastern side. In the north and northwestern part of the basin weather is very cool owing to its altitude. The other parts of the basin except the Lower Palanis enjoy a dry weather.

The rainfall has receded from the hilly terrain to the centre of the valley. The Lower Palanis and the Varushanad hills receive more than 1200 mm of rainfall per annum. The valley receives major part of the rainfall during the northeast monsoon season. Veerapandi receives lowest rainfall (592 mm) in the basin. The central stretch of the basin has less than 700 mm per annum. Most of the plain region receives less than 800 mm of rainfall. As the basin lies on the leeward side of the Cardamom hills, the average rainfall is only about 883 mm per annum. The study area is subject to frequent failure of monsoons. A detailed account of rainfall distribution is dealt with in the third chapter dealing with the evaluation of water resources.

The relative humidity is high during the monsoons and low during the remaining months. The mean annual relative humidity of Kodaikanal, Kamatchipuram, Viralipatti and Madurai are 77.0, 71.2, 59.6 and 77.1 per cent respectively. In general the relative humidity increases from June and attains a maximum in November for then on it decreases to a minimum during the dry months of March and April.

The winds have direct impact on the climate and vegetation of the basin. The mean annual wind velocity of Kodaikanal, Viralipatti and Madurai are 12.9 kmph, 9.7 kmph and 8.7 kmph respectively (Appendix 2.1). The monsoon winds which are active in the west coast from June to September after dropping its moisture on the west coast rush through the gaps of Sakkaluthu, Kathu Kanavoy and Uchaluthu and get deflected by the mountains and spurs of the eastern portion of the ghats and start whirling. Due to the whirling action of the winds at enormous velocity, the soil gets drilled and scooped.

2.2.4 Natural Vegetation

The Upper Vaigai basin is noted for its rich and interesting flora in South India (Hart, 1933). Many new field and plantation crops and trees were introduced over the years. The important ones are coffee, tea, chinchona, pyrethrum, geranium, blue gum, black wattle, conifers, grasses, potatoes, sweet potatoes, tapiaco, maize and cardamom.

The tropical and subtropical climates of the basin sustain all types of vegetation which is one of great advantage for the introduction of various types of plants. Based on the climate, the forests of the basin can be classified into five categories.

The tropical wet evergreen forests are situated on the eastern slopes of the Cardamom hills and the western part of the Varushanad hills (Fig.2.5). Montane wet temperate forest covers the Upper Kottagudi basin. Owing to the very heavy rainfall (Devikolam - 2428 mm) and high altitude this area is called montane wet temperate forest. The subtropical broad leaved hill forests cover the northeastern part, mostly the Lower Palanis where the annual rainfall at Berijam and Kodaikanal are 1437 mm and 1615 mm respectively. A small patch of subtropical broad leaved hill forests are also seen in the Upper Kotagudi basin to the west of Bodinayakannur.

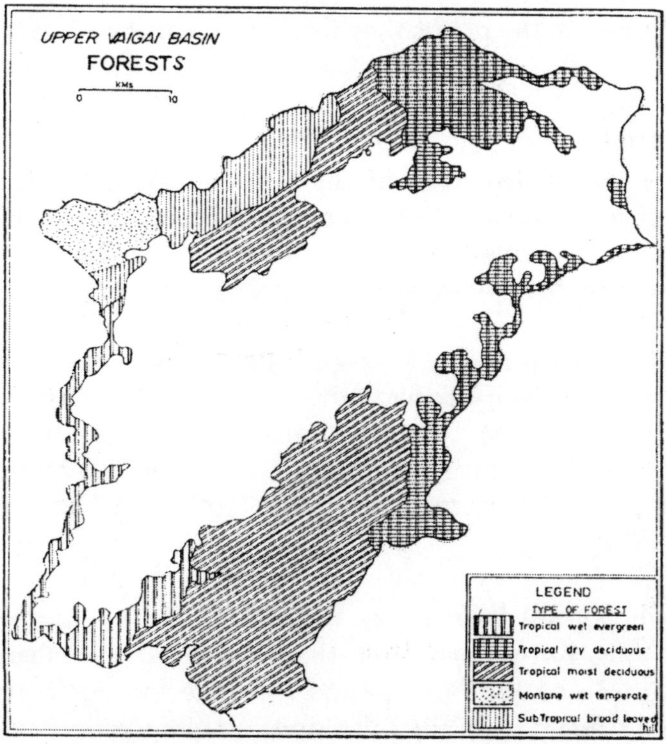

Fig. 2.5

The tropical moist deciduous forests are noticed in the Agamalai, the Kodaikanal hills, the Varushanad and parts of the Andipatti hills. The eastern part of the Lower Palanis and the Andipatti hills are covered with tropical dry deciduous forests.

Thorny plants, shrubs and deciduous trees are spread over the Andipatti hills of the basin. The vegetation of the plains comprises of scattered trees on the fields, avenue trees and shrubs as live fences around the fields and along the sides of cart tracks (Hart, 1933).

The main timber forest is found in the area around Kambam Valley Reserved Forests. The area of these forests is insignificant but this will increase in course of time with protection as forests are very helpful to the people in various ways. It improves the capacity of the soil to absorb and retain rainwater, letting it out gradually by means of springs and also prevent soil erosion. In thickly wooded areas the springs are perennial and are indispensable to the residents of the hills and in the plains they feed the streams and rivers (Hart, 1933).

2.2.5 Drainage

The general direction of the Upper Vaigai River is from southwest to northeast. It is a perennial river. The Suruliar, the Kottagudi, the Varahanadi, the Manjalar and the Maruthanadi are important tributaries of the Vaigai (Fig.2.6). The Upper Vaigai basin is divided into eleven sub-basins, namely the Thalai Vaigai, the Suruliar, the Thevaram, the Kottagudi, the Koilpatti, the Varahanadi, the Manjalar, the Maruthanadi, the Viralipatti, the Viruveedu and the Andipatti (Fig.2.6). Dentritic drainage pattern is noticed in the Thalai Vaigai, Kottagudi, Varahanadi, Manjalar and Maruthanadi whereas parallel pattern of drainage is observed in the Lower Suruliar basin.

2.2.5.1 The Vaigai River

The Vaigai originates from the jungle clad Varushanad hills at an altitude of 1896 m. it runs towards the north receiving numerous feeders from the hill slopes on either side. The general direction of the Thalai Vaigai is from south to north. It is a non-perennial river. Dentritic drainage pattern is noticed in the hilly areas of Varushanad. It emerges from the hilly tract into the plain near Duraiswamipurmam and joins the Suruliar near Vallalnadi (Muthuswamy, 1933). The total area of the Thalai Vaigai basin is about 1095 km^2.

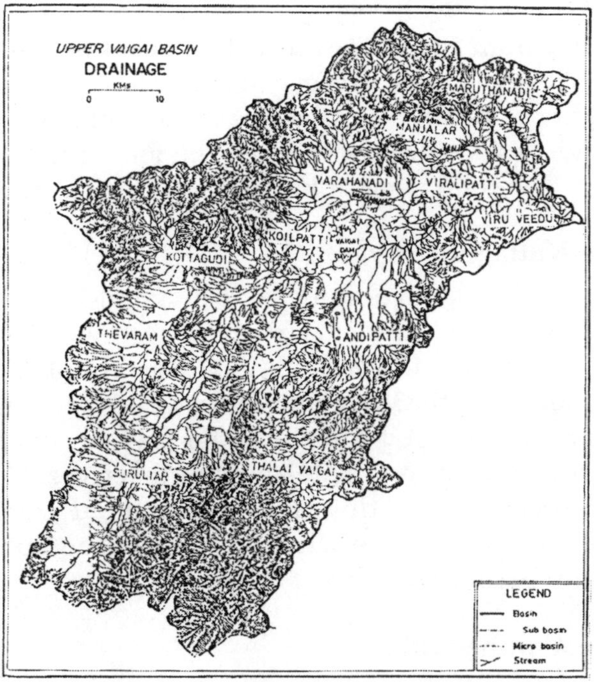

Fig. 2.6

2.2.5.2 The Suruliar

The river Suruli rises in the High Wavy Mountains at an altitude of 1896 m and leaps down in a westerly direction and after a descent of 2 km enters the plains and Suruli falls located here. The general direction of the Suruliar is from southwest to northeast. It is a perennial river. Dentritic drainage pattern is noticed in the Highwavy Mountain and parallel pattern of drainage is observed in the Lower Suruliar basin. It runs through the rich alluvial plains, which is very fertile built-up by the fine soil brought by the Vairavnar and Suruliar (Ayyengar, 1933). The waters from Periyar Lake, situated in Kerala State, are transported to the Suruliar basin through a tunnel. Very few tributaries join this river from the west for they are not in a position to reach the river because of poor water supply. The Evarangalar is an important tributary from the east. Near Theni the river joins the Theniar. It has a series of anicuts all along its course and helps in the supply of water for irrigation. The total area of the Suruliar basin is 695 km^2.

2.2.5.3 The Thevaram River

The Thevaram River rises near Ramakkal in the Cardamom hills of the Western Ghats. The general direction of the river is from southwest to northeast. Near Perumalkarundampatti its direction changes and flows towards the east the joins the Suruliar near Uppukottai. The total geographical area of the basin is 278 km^2.

2.2.5.4 The Kottagudi River

This river rises at an altitude of 2640 m in Misappuli Malai of the Lower Palains and descends from the northwestern part of the hilly region and flows towards the southeast. Dentritic pattern of drainage is noticed in the Kottagudi River. The river enters the plains near Melparavi where it is known as the Theniar. From Agamalai various jungle streams originate and join this river. Near Theni it joins the Suruliar. Since this river is perennial water resource developmental activities are active. The total geographical area of the Kottagudi basin is 475 km^2.

2.2.5.5 The Varahanadi

The Varahanadi, a perennial river, originates from the Lower Palanis at an altitude of 2451 m and flows in a northwest to southeast direction. Dendritic drainage pattern is seen in this area. A tributary, the Periyar, descends from Berijam Lake and joins Varahanadi before entering Periyakulam. The Varahanadi finally joins the Vaigai near Gullapuram. The total area of the basin is 366 km^2.

2.2.5.6 The Manjalar

The Manjalar rises in the Perumalmalai of the Lower Palanis at an altitude of 2234 m. it passes through deep gorges and has a fall of 303 m. the Iruttar and the Varattar join this river on the left and right banks respectively. The Manjalar dam is constructed here. The general direction of the river is northwest to southwest and joins Maruthanadi near Reddiarpatti. The total geographical area of the basin is 308 km^2.

2.2.5.7 The Maruthanadi

The Maruthanadi rises from the Arankanal Malai of the Lower Palains at an altitude of 633 m. Its general direction is from the northwest to southeast. Maruthanadi dam is constructed near Ayyampalayam. After receiving two tributaries from the north its direction changes towards the south. The river joins Manjalar near Reddiarapatti and finally joins the Vaigai River near Ayyampalayam. The total geographical area of this basin is 323 km^2.

A few tributaries join the Vaigai River from the Andipatti hills. The general direction of these tributaries is from south to north. Their maximum length is less than 10 km. The important tributary over here is Marutha Odai.

A dam is constructed across the Vaigai River near Andipatti. The Varattar which descends from Agamalai in the northeastern part of the hilly region drains into the Vaigai dam. In between the Varahanadi and the Manjalar a few tributaries join the Vaigai River from the north.

The river Vaigai flows in an easterly direction from Vaigai dam up to Anaikaraipatti with a broad meandering belt. The direction slightly changes towards the northeast near Anaikaraipatti and from here it traverses a length of about twenty kilometers till it reaches Ayyampalayam. Since the slope is towards the southeast the direction of the river changes from Peranai.

2.2.6 Soils

The soils of the study area are classified based on the Soil Survey and Land Use Organization of the Government of Tamil Nadu (Soil Survey and Land Use Organization, 1980 and 1981). The scheme of classification followed is adopted from soil taxonomy. Three different soil orders such as entisole, vertisoles and alfisoles are generally found in the basin (Fig.2.7).

Entisols (recent soils) are usually found on steep actively eroding slopes and on flood plains that receive new deposits of alluvium at frequent intervals. In this soil order ustorthents group is found in the basin. These are red, non-calcareous, insitu, shallow and well drained soils. It is subjected to severe sheet erosion and often exposing the parent material. Quartz gravels are usually present on the surface of the soil. This has been given the name 'Irugur'.

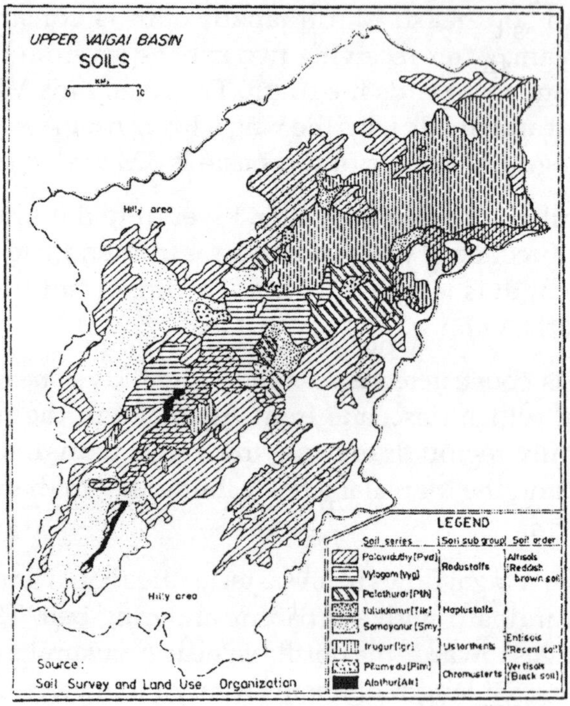

Fig. 2.7

Vertisols (commonly referred to as black soil) are cracking clay soils mostly montomorillinite that shrink and swell over the seasons and that have wide deep cracks during the periods of moisture deficiency. Alkaline reaction is a feature common to the various parent materials like calcareous sedimentary rocks, basic igneous rocks and alluvial deposits. Vertisols are usually formed over gentle slopes. The important groups found in the basin are: Pilamedu and Alathur and these are named after these places. These are black, heavy textured, calcareous, cracking, and very deep and usually have gilgai micro relief and slicken sides. They are moderately well drained with slow permeability ad moderate in reaction.

Alfisols (reddish brown and red sandy soils) are the most extensive soils found in the basin. They have either anargillic or a neutural horizon. The base saturation is more than 35 per cent blow the top argillic horizon. These soils are characterized by large clay accumulation and have usually red or reddish brown colour. The most common group of soils found in the basin is 'Palaviduthi', 'Vylugam' and 'Somayyanur'. The Palaviduthi and Somayyanur series are transported soils whereas Vylogam series sedentary soils and lateritic in origin. Somayyanur series are calcareous whereas other two are non-calcareous. They are deep to very deep, moderately well drain with moderately rapid permeability and natural to moderately alkaline in reaction.

The Somayyanur and Palaviduthi series are of colluvial origin, along the pediments of the Cardamom hills from where the colluvium would have resulted. Presence of cobbles in the profile and deep neutral of the soil reveal that they are of colluvial origin. Somayyanur series are calcareous and carbonates are present at various depths depending upon the deposition and stabilization. The Palaviduthi series are non-calcareous. Palathurai and Tulukkanur series are calcareous on account of having developed from rocks containing calcium carbonate. Porous nature of the surface soil and undulating topography facilitating severe erosion has reduced the soil depth considerably.

The rest of the series occur in the inland plains. Irugur series occur at comparatively higher slopes and are associated with isolated hillocks and is subjected to severe sheet erosion exposing parent material often. The central part is covered by Vylogam series which is a sedentary soil and lateritic in origin. The topography is such that both Vylogam and Irugur series are susceptible to erosion rendering to shallow soils in many places. Sand dunes are found in areas having gentle slopes.

The undulating central plain houses all the other insitu soil series. Pockets of Pilamedu series occupy table lands in the Andipatti sub-basin. The summit is occupied by the shallow and light textured soils, i.e. Irugur series. Slope is occupied by light to medium textured and moderately deep soils i.e. Vylogam, Palathurai and Tulukkanur and the basin is occupied by medium textured and deep to very deep soils. Appendix 2.2 gives the characteristics of different soil series of the study area.

The moisture of the soils ranges between 0.4 and 5.8 per cent. Irugur soils being light textured recorded the minimum moisture i.e., less than one per cent. Pilamedu and Sommayyanur which are heavy textured with moisture ranging between 2 and 58 per cent. The water holding capacity is high in Pilamedu and Somayyanur series ranging between 50 and 65 per cent and is medium in other zones ranging 34 to 47 per cent. It is low in san dunes, high in Pilamedu and in between in other series of the soils.

The Kombai and adjacent areas of Bodinayakkanur are prone to sand dunes. Soil erosion and consequent surface runoff pose serious problem in the Thevaram sub-basin. The severely affected soils in the basin are Irugur and Vylogam series. In order to overcome these difficulties soil conservation measures are being taken up in the basin.

2.2.7 Geomorphology

The geomorphic zones of the basin are depicted in Fig. 2.8. Varius geomorphic zones are identified with the help of landsat images using false colour composites. The tonal variations, texture and alignment parallel to the hilly terrain are made use in deciphering the zone. The six geomorphic zones of the Upper Vaigai basin are as follows: hilly terrain, bajada, and valley filled sediments, sand dunes, pediplain and flood plains.

The hilly terrain bordering the basin shows 'ridges and valley' topography. The formation of valleys and ridges are mainly controlled by either the structural features of the rock formation or lineaments. There is an extensive upland valley called Varushanad Valley carved by the Thalai Vaigai River at the top of the ghats. The remnants of ridges are found here and there. At most of the places the periphery of the ghats is marked by a precipitous scarp. The scarp slope is dissected by 'V' shaped valleys with steep gradient and several waterfalls and rapids. The Suruli waterfall is worthy of mention.

Bajada zones are seen after the hilly terrain in almost all the western, northern and eastern sides of the basin. These zones are characterized with exposed barren rock out crops near Andipatti hills on the northeastern part of the basin. Geophysical studies reveal that the thickness of sediments on the southwestern part of the valley ranges between 20 m to 30 m (Public Works Department, 1976). The groundwater potential is more in these areas.

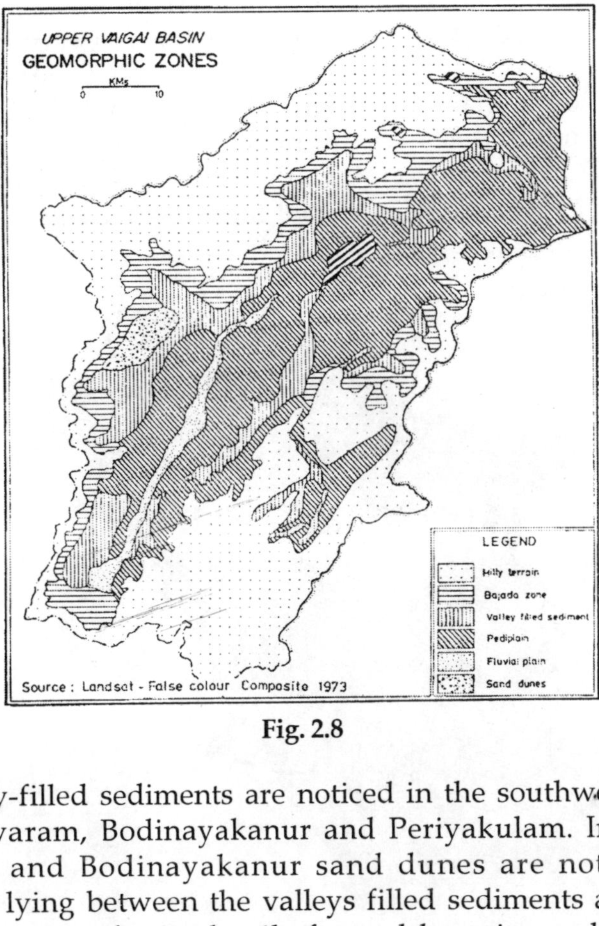

Fig. 2.8

Valley-filled sediments are noticed in the southwestern part near Thevaram, Bodinayakanur and Periyakulam. In between Thevarm and Bodinayakanur sand dunes are noticed. The pediplain lying between the valleys filled sediments and fluvial plain comprise of mixed soils formed by rain-wash from the bordering valley filled sediments. The fluvial plains on both sides of the Suruliar are filled by the materials deposited by the Suruliar on either side of the banks during the peak flood periods. These are composed of rich alluvial soils. In a few pockets along the hilly terrain particularly near Bodinayakanur colluvial fans are deposited by the action of streams.

2.2.8 Relative Relief and Average Slope

The relative relief of the basin is worked out based on Smith's method. (Monkhouse and Wilkinson, 1967) and depicted in Fig.2.9. It gradually decreases in the south and southeastern slopes of the

Lower Palains and is rather pronounced in the west of Thevaram. i.e. more than 750 m. The amplitude of relief is very high in the western portion of Varushanad hills, whereas in the eastern side it is less than 300 m. In the Varushanad Valley the relative relief is less than 150 m and in the Andipatti hills the relative relief is between 150 and 450 m. Apart from this, the entire valley region has less than 150 m of relative relief except for a few packets.

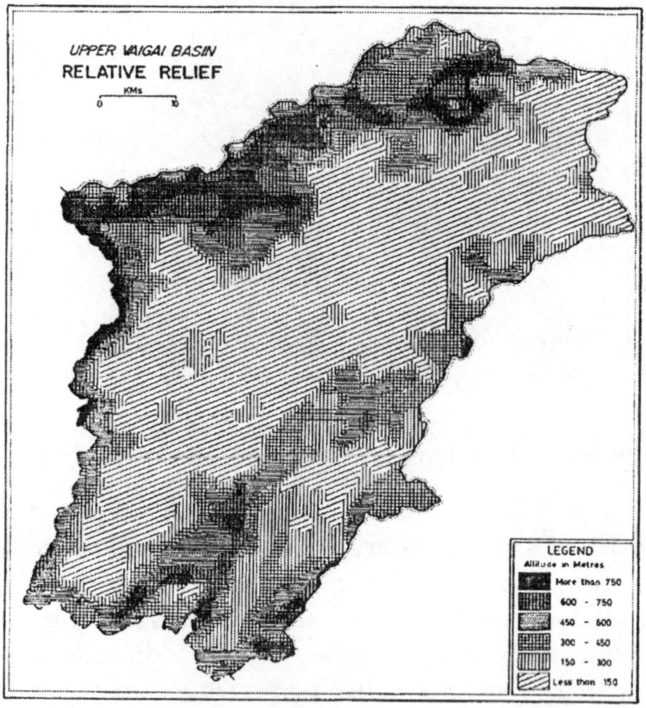

Fig. 2.9

A slope analysis is attempted for the basin adopting the method developed by Wentworth (Fig. 2.10). The slope varies from less than 1° to over 30°. Land is almost gentle in about 17 per cent of the area of the basin and this is mostly found in the areas of the basin drained by the Thevaram, the Lower Kottagudi, the Lower Maruthanadi and the Lower Varahanadi. Slope is found to increase rapidly with elevation. The central stretch of the basin where the average slope is less than 1° is dotted with areas of steep slopes in the vicinity of the isolated hillocks (Fig. 2.10). In the plateau region the slope varies from 1° to 3° and here again one finds that the plateau in the vicinity of the Upper Suruliar and Thevaram is bordered with a more or less undulating topography. In the rest of the plateau region the slopes vary rather sharply towards the highland rim. Beyond the plateau the slope increases between 5° to over 30° and the areas of maximum slope are found along the crest of the Lower Palanis.

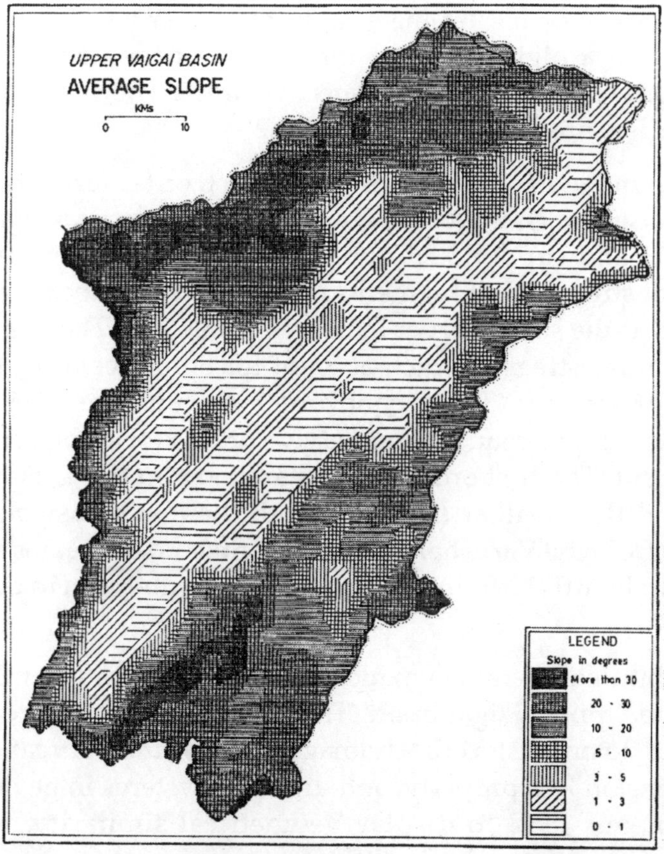

Fig. 2.10

2.2.9 Drainage Morphometry

The salient features of the drainage morphometry which includes stream orders, stream number, stream lengths and stream density have important bearing on the water resources of the basin.

2.2.9.1 Stream Orders and Stream Numbers

The Upper Vaigai River is designated as eighth order stream according to the Strahler's system. There are two seventh order streams namely, the Suruliar and the Thalai Vaigai. The Kottagudi River is designated as the sixth order stream, the Varahanadi (6) Manjalar (6) Maruthanadi (6) and Viralipatti (4). These rivers directly drain into the eighth order stream without forming the intermediate orders of streams. This is a drainage peculiarity, which may be due to the presence of harder material in the lower part of the basin, which retard a high degree of integration of the fifth, sixth and seventh order streams in the respective areas. The sub-basin-wise stream orders are presented in Table 2.2.

The number of stream segments of each order form an inverse geometric sequence with order number in the Upper Vaigai basin (Table 2.2) and supports Horton's law of stream order. The total number of stream in all segments is 18016. The number of streams decreases as the stream order increases (Table 2.2). There are 14416 the first order streams which account for 80 per cent to the total number of streams. The second order streams are 2746, which contributes 15 per cent. The successive orders form the remaining five per cent. The first order streams are more in the Thalai Vaigai (4875) and the Suruliar (2766), as both these sub-basins streams originate from the Varushanad hills. The first order streams are very few in Viralipatti (120) since it is the smallest sub-basin in the study area.

The bifurcation ratios characteristically range between 3.75 and 5.25 in the Upper Vaigai basin (Table 2.2) in which the geological structures do not distort the drainage. Because the bifurcation ratio is a dimension less property, and drainage systems in geologically homogeneous tend to display geometrical similarity, it is not surprising that the ratios in the Upper Vaigai basin show only a small variation from region to region. The regression coefficient is identical with the logarithm of bifurcation ratio is estimated to be 4.07 which means that, on an average, there are about four times as many channel segments as of any given order in the basin (Fig. 2.11A).

Table 2.2
Upper Vaigai Basin
Morphometric Characteristics

Name of the sub-basin	Area in km2 - Au	Stream order - u	Total number of streams - Nu	Bifurcation Ratio - Rb	Total Length of Streams in Km - L	Length Ratio - RL	Perimeter in Km - P	Maximum basin length - Lb	Stream Density - D	Stream Frequency - F	Diameter of Basin in Km-d	Form Factor - Rf	Circulatory Ratio - Rc	Elongation Ratio - Re	Constant of Channel maintenance - e	Length of overland flow - Le
Upper Vaigai Basin	3731.74	8	18016	4.07	8407.05	2.123	335.00	103.00	2.25	4.83	68.93	0.3518	0.4179	0.6622	0.5099	0.2550
Thalai Vaigai	842.16	7	6030	4.30	2565.50	3.045	169.00	55.00	3.05	7.16	32.75	0.2784	0.3705	0.5955	0.3279	0.1639
Suruliar	696.16	7	3389	3.92	1392.50	2.742	162.50	57.00	2.00	4.87	29.77	0.2143	0.3313	0.5222	0.5000	0.2500
Thevaram	261.87	5	667	5.15	445.00	3.168	69.00	26.00	1.70	2.55	18.26	0.3874	0.6912	0.7023	0.5882	0.2941
Kottagudi	402.71	6	2252	4.59	978.25	1.660	96.25	34.25	2.43	5.59	22.64	0.3433	0.5463	0.6610	0.4115	0.2058
Koilpatti	136.29	8	404	3.75	225.00	1.758	66.50	18.75	1.65	2.96	13.17	0.3877	0.3873	0.7024	0.6061	0.3030
Varahanadi	419.52	8	2037	4.42	996.25	3.300	100.25	31.50	2.37	4.86	23.11	0.4228	0.5246	0.7337	0.4219	0.2110
Manjalar	264.69	8	1127	4.10	534.50	2.508	91.50	32.00	2.02	4.26	18.36	0.2585	0.3973	0.5736	0.4950	0.2475
Maruthanadi	244.90	8	840	5.24	466.00	4.464	103.50	36.25	1.90	3.43	17.66	0.1864	0.2873	0.4872	0.5263	0.2362
Viruveedu	148.03	5	478	4.50	288.00	1.708	65.00	24.25	1.95	3.23	13.73	0.2517	0.4403	0.5662	0.5128	0.2564
Viralipatti	66.74	8	167	4.01	109.50	2.190	23.00	11.75	1.64	2.50	9.22	0.4834	1.5853	0.7847	0.6098	0.3049
Andipatti	248.64	5	629	4.79	407.00	2.580	63.00	25.50	1.64	2.53	17.79	0.3824	0.7873	0.6976	0.6098	0.3049

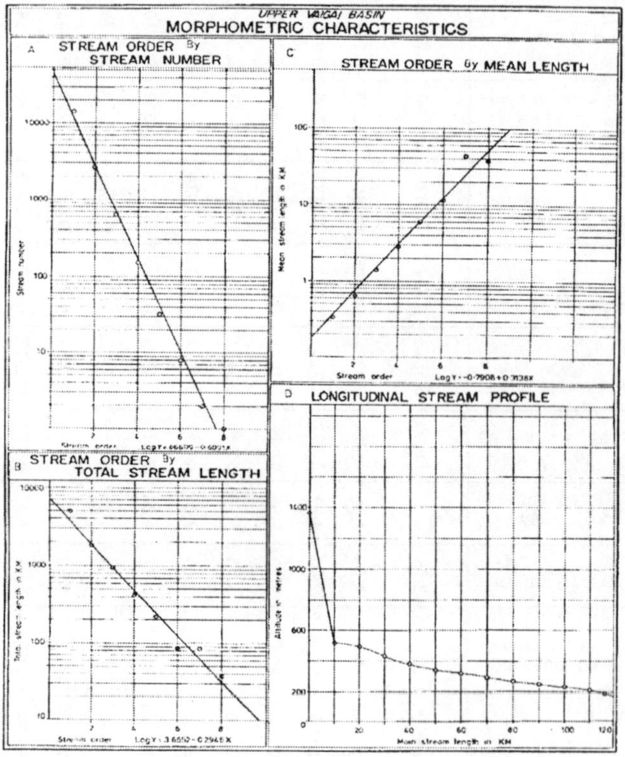

Fig. 2.11

2.2.9.2 Stream Length

The total length of all streams of the Upper Vaigai basin is 8407.50 km, of which, first order streams have a length of 4856 km, which accounts for 58 per cent of the total length of all orders. The second order streams have a length of 1764 km accounting 21 per cent to the total stream length in the basin. Both the seventh order streams the Suruliar and the Thalai Vaigai, have a length of 88.75 km together and the eighth order stream, the are plotted against the stream orders on a semi-logarithmic paper and the plot yields a linear relationship (Fig.2.11B).

The length ratio, which is the ratio of the mean length of the streams of a given order to the mean length of the streams of the next order, is then computed for each pair of orders for the Upper Vaigai and its sub-basins. The total length, mean length, the length ratio and computed mean ratios are presented in Table 2.2. The length ratio for different order of streams varies from 0.85 to 4.00. Very high length ratio is noticed in the seventh order streams and the lowest is found in the eighth order stream. The length ratio is very high in the seventh order streams of the Thalai Vaigai and the Suruliar (9.0 and 7.2 respectively). In the Thevaram basin also the length ratio of the fifth order stream is 6.52 whereas in the other basins it is very low for different orders of streams. The mean lengths are plotted against the stream orders on a semi logarithmic paper and the plot yields a linear relationship (Fig.2.11C).

Verification of Horton's law of stream number and lengths support the theory that geometrical similarity is preserved in this Upper Vaigai basin of increasing order (Fig. 2.11C).

2.2.9.3 Drainage Density and Stream Frequency

Drainage density is the ratio of total channel-segment length cumulated for all orders within a basin to the basin area. Dimensionally, this ratio reduces to the inverse of length. The Upper Vaigai basin is found to have a drainage density of 2.25 km/km^2 (Table 2.2) The drainage density is more in the Thalai Vaigai basin (3.05) followed by Kottagudi (2.43), the Varahanadi (2.37) the Manjalar (2.02) and the Suruliar (2.00). In other sub-basins of the study area the drainage density is less than 2.00 km/km^2.

The Upper Vaigai has a stream frequency of 4.83. The stream frequency is more in the Thalai Vaigai basin (7.16). In the Kottagudi, the Suruliar and the Varahanadi sub-basins the stream frequency is 5.59, 4.87 and 4.87 respectively. In other sub-basins the stream frequency is less than 3.5.

2.3 CULTURAL BACKGROUND

2.3.1 Population Characteristics

The Upper Vaigai basin contains 129 villages and 11 towns. The total population in 2001 was 1089785, of which about one third lives in rural areas and the rest in urban centers.

People are mostly distributed in the river valleys while the highlands shun population and settlement. The only large urban centre in the highlands is Kodaikanal, a summer resort, located in the Lower Palanis to the north. Population growth in this basin has been phenomenal since the construction of the Periyar Irrigation project which has converted the thirsty lands of the Upper Vaigai basin into a region of prosperity and hope.

2.3.1.1 Distribution of Population

The distribution of population is depicted in Fig. 2.12 as in 2001 and it reveals a very uneven pattern. Areas along the rivers Suruliar, Kottagudi, Varahanadi and Manjalar have the highest concentration of population. The highest concentration of population in this region could be attributed for the better irrigation facilities, fertile soil and other better natural conditions suitable for growing food and cash crops. Distribution, however, gets thinner in the Thevaram, the Andipatti, the Viruveedu and the Varahanadi sub-basins. These are the less productive patches of the basin due to the absence of assured irrigation facilities, poor soil, uneven topography, poor accessibility and hence thin population. The Varushanad-Andipatti hills, the Cardamom and Lower Palanis have no population at all except a few pockets like Kodaikanal, Vellagavi, Adukkam, Pannaikadu, Kamanur and Agamalai in the Lower Palanis and Myladumparai and Kadamalaikundu in the Varushanad hills.

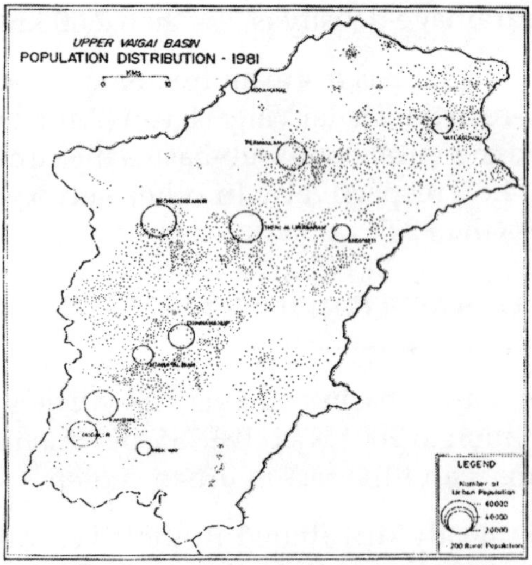

Fig. 2.12

2.3.1.2 Proportion of Urban Population to the Total Population

The urban population, which accounts for 32.74 per cent of the total population, resides in the eleven urban centers of the basin. Only five sub-basins namely, the Suruliar, the Kottagudi, the Varahanadi, the Manjalar and the Andipatti have urban centers. The Kottagudi sub-basin with Bodinayakkanur and Theni-Allinagarm as the two urban centers contains 74.6 per cent of the urban population of the area followed by the Suruliar (49.91 per cent), the Varahanadi (43.17 per cent), the Manjalar (25.03 per cent) and the Andipatti (23.44 per cent).

2.3.1.3 Density of Population

Fig. 2.13 depicts the regional variation in the density of population per square kilometer in the basin as in 2001. The density of population is strikingly uneven. Though the overall density of population in the basin is 274 persons per km^2 some parts record a high density of more than 2000 persons per km^2 while some other parts record less than 10 persons per km^2.

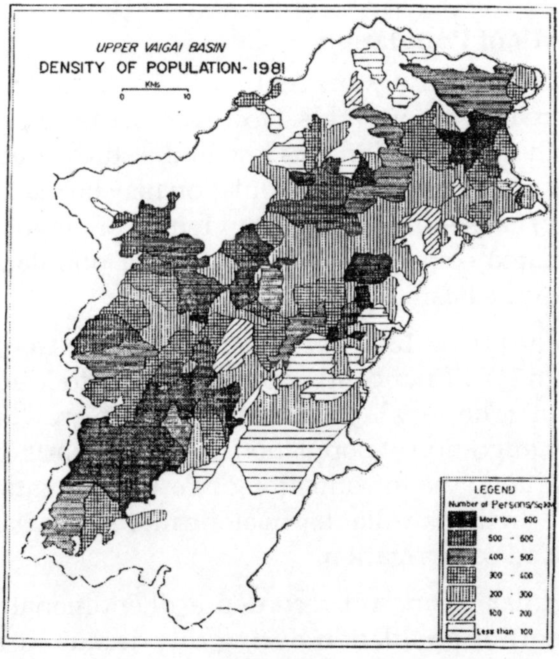

Fig. 2.13

Very high density of population of more than 600 persons per km^2 is found in the Suruliar, the Varahanadi, the Manjalar and the Andipatti sub-basins. These areas are mainly benefited by canals and well-served by roads. Owing to the better irrigation facilities and fertile soil this region forms the best agricultural areas of the basin supporting a very high population.

More than 40 per cent of the area has high density of population of 200 -400 persons per km^2 comprising of the Lower Varahanadi, the Thevaram, the southern most villages of Suruliar and the Upper Maruthanadi sub-basins.

The mountain regions of the study area have low densities of less than 10 persons per km^2. This is mainly because of high relief, steep slopes and thick forest cover and consequently poor soil cover and unorganized irrigation facilities, soil erosion and poor and primitive agricultural practices and lack of transportation facilities. A few villages along the river course have comparatively low population density because of frequent floods and lack of good agricultural lands.

2.3.1.4 Growth of Population

It is found tat during this century the population of the basin has been in creased to four folds. However, in the years 1911-21, the rate of growth of population was low in the study area i.e. 7.75 per cent. The low rate of increase in population may be mainly attributed to a number of causes like draught, epidemics of plague and malaria which decimated considerable proportion of population living in many parts of the basin during the period.

However, during 1891-1901, the rate of growth of population was very high (17.71 per cent) when compared to Madurai district (10.20 per cent). The very high rate of population growth in the basin was due to immigration of population from other parts of the country to the basin due to the introduction of Periyar Irrigation Project in the year 1895. In the Suruliar basin alone more than 5900 ha of land was introduced for irrigation.

The Suruliar basin had recorded a phenomenal growth rate (33.72 per cent) of population during this decade. It appears that the 'pull' factors of employment opportunities in this area due to enhanced supply of water for irrigation stipulated migration of population in substantial number fro other parts of the country in search of gainful employment.

In the year 1971 high rate of growth of population (24.52 per cent) occurred in the basin. This occurred perhaps, due to the competition of different minor irrigation projects like Manjalar dam, Varahanadi etc., in the study area which attracted new settlers in large numbers. The decline in mortality rate also responsible for the high rate of growth.

Owing to the increasing population pressure in the basin there exists an acute need of water for agriculture, domestic and industrial purposes.

2.3.2 Agriculture

2.3.2.1 Land Use

Fig. 2.14 depicts the land use of the study area based on the IRS 1C images. Spatial distribution of different land uses, such as, forests, barren rock, out-crops, reservoirs and agricultural regions are identified. The Lower Palanis, the Cardamom hills and the Varushanad-Andipatti hills are covered with forests with a few patches of barren rock out-crops here and there on these hilly regions. A strip of land along the Suruliar course in intensely utilized for crops because of canal irrigation practiced here. The tank and well irrigated areas are found in the Thevaram, the Kottagudi, and the Varahanadi, the Manjalar, the Aruthanadi and the Andipatti sub-basins apart from the rain-fed crops identified in the other areas of the basin.

The total geographical area of the study area is 3976.35 km^2. Forest covers an extent of 37.1 per cent to the total area followed by net sown area 36.2 per cent, land under miscellaneous use (10.0 per cent), current follows (9.0 per cent), uncultivable land (5.6 per cent), cultivable waste (1.2 per cent), other fallows (0.6 per cent) and grazing lands and land under miscellaneous tree crops (0.2 per cent each).

2.3.2.2 Irrigation

The Upper Vaigai basin noted for the frequent failure of monsoons. The sub-soil is thin and the slope is to steep in most

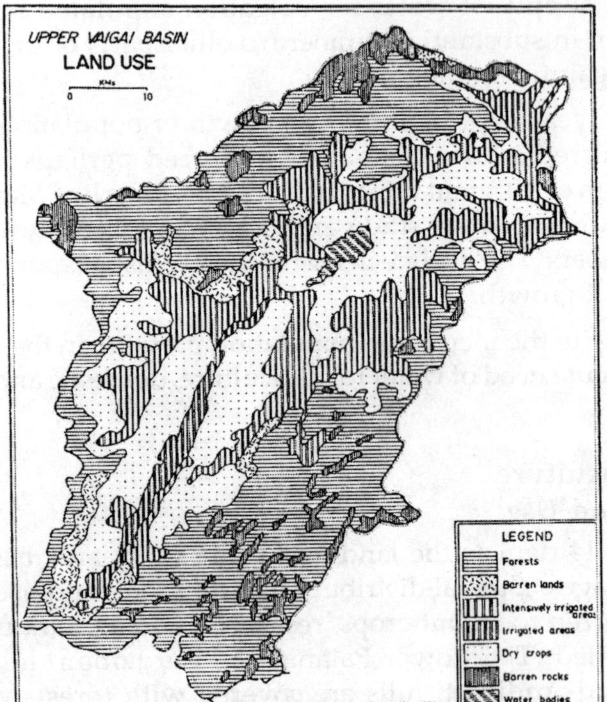

Fig. 2.14

parts of the basin and hence irrigation is restricted to the low lands drained by the rivers. However, due to the Periyar irrigation Project, the Kambam Valley has been completely irrigated.

The area under study has different sources of irrigation. The chief sources in the order of importance are the canals, tanks and wells. Canal, the largest source of irrigation, is confined to the Suruliar, the Manjalar and the Manuthanadi sub-basins. The rain-fed tanks and wells which form other sources of irrigation are also found to exist in almost all the sub-basins of the Upper Vaigai basin.

Canal irrigation covers more than 17 per cent of the total irrigated area. The sources of irrigation are the Periyar Vaigai canals, the Manjalar and the Maruthanadi canals. The river has been fully exploited by means of several anicuts both old and new. Apart from this, there are a few jungle streams but owing to their capricious and violent behavior they are not of much use.

Table 2.5
Upper Vaigai Basin
Mean Land Use – 1999-2002

Name of the sub-basin	Total area	Percentage to total area								
		Forests	Uncultivable Land	Land under miscellaneous use	Cultivable Waste	Grazing Lands	Land under miscella neous Tree Crops	Current fallow	Other fallow	Net sown area
Thalai Vaigai	109509	46.4	4.7	17.0	0.8	0.4	0.1	5.1	0.2	2.5.3
Suruliar	69536	34.7	3.3	6.4	1.0	-	0.3	10.9	1.1	42.2
Thevaram	27837	13.3	10.2	9.6	1.8	0.1	-	12.6	1.5	50.9
Kottagudi	47498	57.3	1.3	4.2	0.8	-	-	7.2	0.1	29.1
Koilpatti	8441	-	17.4	19.3	1.0	-	-	10.2	1.9	49.9
Varahanadi	36555	23.8	6.5	12.1	2.0	0.1	0.5	9.4	0.6	45.0
Manjalar	30754	51.1	5.0	3.4	1.4	0.3	0.6	7.2	0.5	30.5
Maruthanadi	32274	39.2	14.7	3.7	1.4	-	-	4.8	0.6	35.5
Viralipatti	3761	-	3.5	8.6	1.8	-	-3	45.8	-	40.3
Viruveedu	13033	6.7	5.6	10.1	2.9	-	-	26.3	-	48.4
Andipatti	18437	21.1	2.0	11.2	1.1	0.1	0.7	13.7	-	52.1
Upper Vaigai	397635	67.1	5.6	10.0	1.2	0.2	0.2	9.0	0.6	36.21

The principal source of water for the Periyar Irrigation project is Periyar Lake located in the neighboring Periyar basin of Kerala State. It was originally constructed in 1895. In view of the enhanced supply of water to the fields various anicuts have been constructed all along the Suruliar.

The basin is served by numerous tanks. These tanks can be classified into two kinds such as rain-fed and canal-fed tanks. All along the Suruliar, tanks are connected with canals which supply water to feed the nearby agricultural fields. P.T. Rajan channel, which starts from the Suruliar connects a few tanks and irrigates an area of 543 ha. The rain-fed tanks are more or less isolated and irrigation based solely on these tanks has seldom been successful. Tank provides irrigation for about 12 per cent of the total irrigated area.

Area under well irrigation account for 15 per cent of the total irrigated area in the basin. It is believed that even during the famines of 1877-78 a good number of wells in the region remained a source of relief to the riots.

Though at present canal irrigation has improved the situation greatly, still well could easily be called as the chief source of irrigation for farmers of the non-project area villages. The immense and incalculable value of these could be understood from the fact that in spite of increased irrigation facilities provided by the Periyar Irrigation Project the number of wells have always been on the increase and their depths increasing thus indicating a falling water table The topography of the region also favors well irrigation. Wells vary in depth and are found even in the canal irrigated areas where they function as a supplementary source of irrigation in times of failure of the monsoons. They are the main source of irrigation for the second and third crops. Although bullock driven methods were used for lifting well water until recent times they have now been widely replaced by electric pumps.

CHAPTER 3

EVALUATION OF SURFACE AND GROUND WATER RESOURCES

3.1 INTRODUCTION

One of the major objectives of inquiry of water resources is to establish temporal and spatial patterns of availability and the quantity of the surface and groundwater resources and predict the yield of water at specific places and times. It is of utmost importance to make an appraisal of this resource in order to achieve optimum benefits for the society. Assessment of this sort involves exploration which helps in the development and management of this resource (Sharma, 1971). Water is a dynamic resource and its occurrence and distribution is not predictable in a deterministic manner. Despite the variations in the availability of water from day to day and season to season, the quantum of water available should be estimated with required accuracy and desired confidence.

An attempt has been made in this chapter to assess the quantum of surface and groundwater resources in the Upper Vaigai basin. The surface and groundwater resources are assessed using a number of techniques of water resources evaluation. Employing the book-keeping procedure of Thornthwaite and Mather (1955), water balance is estimated and areas of water deficiency and surplus are identified. The river flow data are assessed using log-normal and Gumbel distribution for the annual flows and central tendency, deviations and flow duration curves for the monthly flows. The groundwater yield is estimated through specific yield and rainfall recharge methods. Trend surface analysis is used to study the overall trend in mean annual rainfall distribution and groundwater table. Apart from this the characteristics of precipitation, groundwater table and water are also dealt with in this chapter.

3.2 SURFACE WATER RESOURCES

Surface water resources remain the most important source of water or the basin requirements and therefore its assessment is a basic requirement for proper planning of water resources development projects (Rao et al., 1971). A dependable assessment of water resources of a given basin can be made by the analysis of observed hydrological and meteorological data and the existing quantum of utilization of river waters. As such, this is achieved by analyzing characteristics of rainfall and river flow and surface water availability and its quality.

3.2.1 Precipitation Characteristics

Rainfall data for twenty four rain gauging stations (Fig.3.1) in and around the Upper Vaigai basin have been collected for 35 to 82 years according to the availability of data. The rain gauging stations and mean seasonal and annual rainfall of these stations are presented in Table 3.1.

3.2.1.1 Mean Annual Rainfall

The Upper Vaigai basin in characterized by sharp areal differences in the quantum of rainfall (Fig.3.2). It is bordered by highlands all around except in the northeast. The central stretch of the basin has a southwest to northeast alignment and the distribution of rainfall very closely follows this alignment. The southwest and northeast monsoons are the two main seasons of rainfall; however, the northeast monsoon accounts for larger share in the total annual rainfall.

The Lower Palanis in the north and northwest and the Varushanad hills in the southeastern corner of the basin receive maximum rainfall (Fig.3.2). In fact, these highlands virtually shelter the central stretch of the basin from the full vehemence of the southwest monsoon winds thereby depriving the region for the heavy rains of this season. Rainfall sharply decreases from over 2000

UPPER VAIGAI BASIN
RAINFALL AND RIVER FLOW
GAUGING STATIONS
KMs
0
10
Manjalar Dam
Vaigai Dam
Veerapandy
Kombam
Gudalur
Power house
Wattrap
LEGEND
Gauging station
Rainfall
River flow

Fig. 3.1

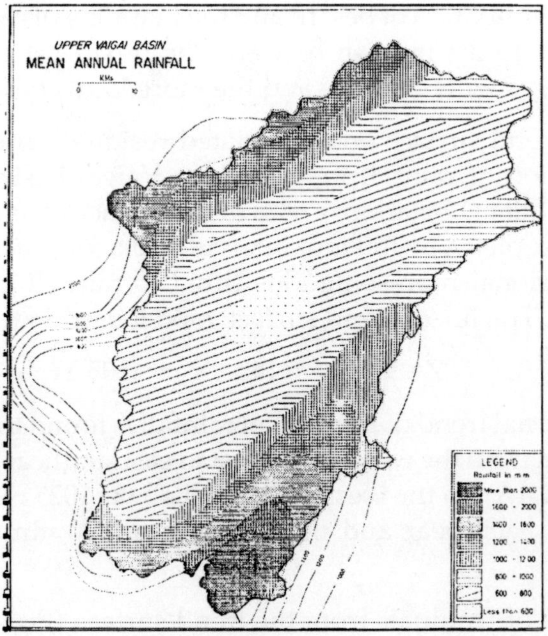
UPPER VAIGAI BASIN
MEAN ANNUAL RAINFALL
KMs
0
10
LEGEND
Rainfall in mm
More than 2000
1600 - 2000
1400 - 1600
Less than 600

Fig. 3.2

mm along the slopes of these hills to about 700 to 900 mm along of foot-hills zones of the lee side of these hills Gudalur, (762 mm), Bodinayakanur (793 mm) and Periyakulam (842 mm), and then to the central stretch of the basin. The rainfall in the central stretch of the basin is less than 700 mm (Kambam, 677 mm; Veerapandi, 592 mm; Vaigai Dam, 640 mm; and Devadanapatti, 685 mm). Thus, the central stretch of the basin forms a dry belt extending from Kambam and Gudalur in the southwest to Devadanapatti in the northeast (Fig.3.2).

3.2.1.2 Mean Annual Rainfall: Trend Surface

Trend surface analysis is adopted to indicate the general distributional pattern of mean annual rainfall. The potential application of trend surface analysis in determining mean annual rainfall is investigated by the Institute of Hydrology (1971) and Chidley and Keys (1970). It attempts to decompose each observation on a spatially distributed variable into a component associated with a regional trend present in the data and a component associated with purely local effects (Unwin, 1975 and Robinson and Bryson, 1957). The computation of best fit and the mapping of residuals from it may enable to distinguish between the broad pattern in which rainfall changes with distance and the more local changes.

Linear trend surface and associated residuals are defined for the distribution of rainfall in the Upper Vaigai basin; taking the station-wise date (17 stations). The trend surfaces are presented in the form of isopleths of 100 mm interval and a graphic portrayal of three dimensional representations fitted to rainfall in the Upper Vaigai basin. The linear trend surface model used in this study is

$$Z = 9.762 - 0.145\ X + 0.03\ Y.$$

The regional trend of the annual rainfall is found to be a surface sloping down from the west-northwest to east southeast of the basin (Fig.3.3 and 3.4A). In the west-northwest about 1025 mm of rainfall are expected each year and this declines to 700 mm in the east-southeast.

The map of residuals from the trend surface (Fig. 3.4B) shows that the highlands of the Lower Palanis and the Varushanad hills

Table 3.1
Upper Vaigai Basin; Mean Seasonal and Annual Rainfall

Rain Gauge Station	Winter		Summer		Southwest monsoon		Northeast monsoon		Mean Annual
	in mm	in per cent	in mm	in per cent	in mm	in per cent	in mm	in per cent	in mm
Gudalur	24.32	3.19	190.53	25.00	244.14	32.03	303.24	39.78	762.23
Kambam	6.75	1.00	131.70	19.45	233.88	34.54	304.71	45.01	677.04
Uthampalaya,	32.85	4.63	146.04	20.16	172.30	24.31	357.56	50.45	708.75
Veerapandi	34.74	5.86	142.63	24.08	102.72	17.34	312.34	52.72	592.43
Rasingapuram	6.47	1.05	100.78	16.34	241.98	39.24	267.41	43.37	616.64
Bodinayakkanur	45.13	5.69	190.03	23.95	167.07	21.06	391.19	49.30	793.42
Berijam	62.01	4.32	288.68	20.10	476.23	33.15	609.59	42.43	1436.51
Kodaikanal	94.51	5.85	335.55	20.78	558.76	34.60	626.18	38.77	1615.00
Periyakulam	63.42	7.53	210.02	24.93	171.87	20.40	397.03	47.14	842.34
Kamatchpuram	20.42	2.76	147.70	20.00	196.10	26.55	374.47	50.69	738.69
Vaigai Dam	18.40	2.87	123.45	19.28	161.00	25.14	337.59	52.71	640.44
Manjalar Dam	14.51	1.98	187.63	25.55	195.15	26.58	336.95	45.89	734.24

Devadanapatti	37.60	5.49	135.00	19.70	194.05	28.32	318.58	46.49	685.23
Maruthanadi Dam	1.50	0.16	246.63	26.90	277.58	30.28	391.15	42.66	916.86
Peranai	24.35	2.67	199.59	21.91	280.85	30.83	406.21	44.59	911.00
Viralipatti	13.68	1.80	158.46	20.90	220.58	29.10	365.36	48.20	758.08
Kumili	60.58	3.85	230.12	14.62	826.08	52.47	457.57	29.06	1574.35
Vandamettu	40.57	1.99	280.35	13.76	1300.01	63.79	417.03	20.46	2037.96
Chinnar	43.40	6.69	88.40	13.62	162.90	25.09	354.50	54.60	649.20
Devikolam	19.54	0.80	353.03	14.54	1668.79	68.73	386.51	15.93	2427.87
Nilakottai	25.99	3.29	160.15	20.25	251.87	31.85	352.70	44.61	790.70
Usilampatti	20.99	2.50	162.98	19.40	220.05	26.19	436.20	51.91	840.22
Watrap	35.88	3.92	169.40	18.52	214.23	23.42	495.22	54.14	914.73
Srivilliputhur	32.93	4.05	187.89	23.12	149.82	18.44	441.89	46.39	812.53

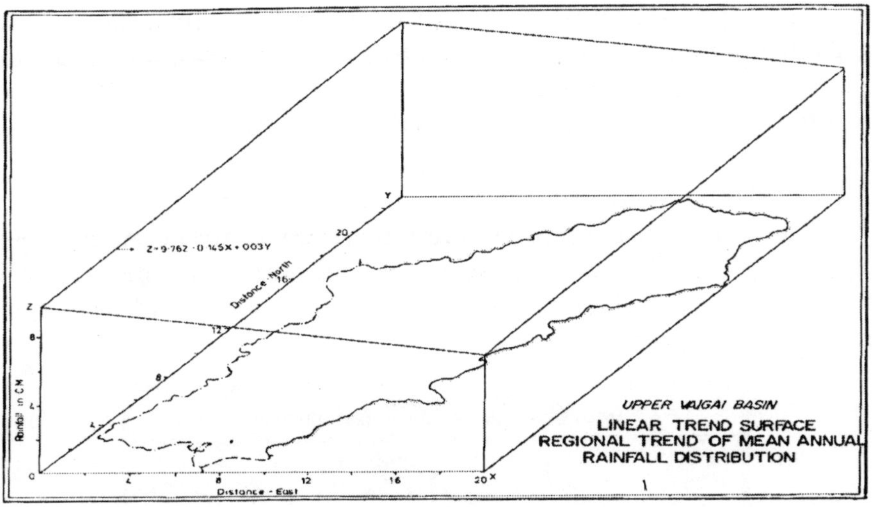

Fig. 3.3

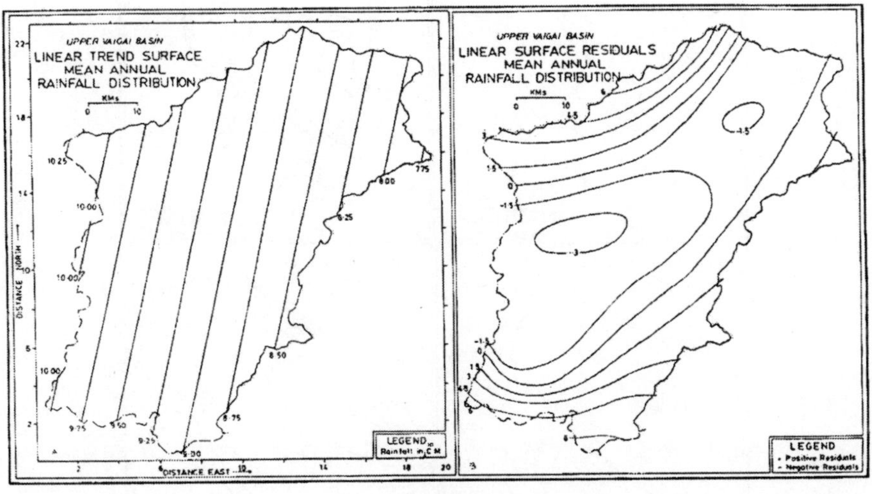

Fig. 3.4

standout as areas with positive residuals whereas the central stretch of the basin are the areas of negative residuals. From the residuals of trend surface it is inferred that in the areas of the Lower Palanis and the Varushanad hills the predicted rainfall surface almost tally with actual surface of annual rainfall distribution. However in the central stretch the predicted surface has a deviation of 1.5 cm from the actual. The isopleths in Fig. 3.4B suggest that the residuals follow a distinct pattern with a ridge of positive residuals towards the Lower Palanis and the Varushanad hills and the maximum negative residuals around Rasingapuram and Periyakulam.

3.2.1.3 Mean Seasonal Rainfall

The spatial distribution of rainfall is seen for four conventional seasons (Fig. 3.5). Winter (January and February), summer (March to May), southwest monsoon (June to September), and northeast monsoon (October to December).

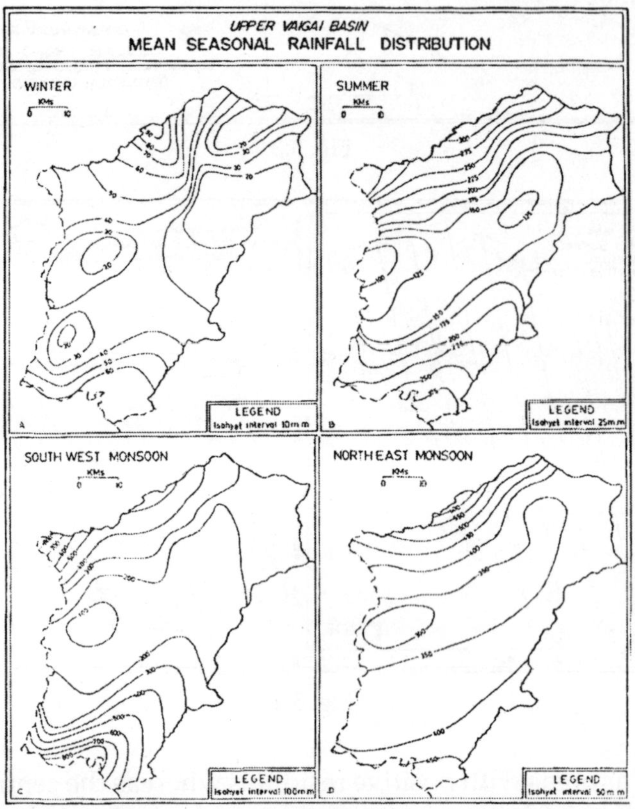

Fig. 3.5

The distribution of rainfall during the southwest monsoon period obviously decreases from southwest to northeast of the basin. The southwest monsoon strikes the Western Ghats and sheds most of the moisture on the windward side resulting in copious rainfall while the leeside of the Western Ghats experiences a very sharp drop in rainfall. For instance, Vandamettu and Kumili on the windward side of the Western Ghats (outside the study area) gets a rainfall of about 2038 mm and 1574 mm respectively while stations like Rasingapuram, Kambam, and Veerapandi just about 30 km to the west on the leeside of the Western Ghats witness a sharp drop in rainfall, less than 600 mm.

The Varushanad and the Cardamom hills in the southeast which run roughly parallel to one another enclosing the southern extremity of the Upper Vaigai basin receive the maximum rainfall during the southwest monsoon period. In the Lower Palanis also rainfall is appreciable during this season. The central stretch of the basin on the other hand gets rainfall of less than 300 mm, which amounts to 35 per cent of the total rainfall during this season (Fig. 3.5).

With the short spell of dry weather in early September, the northeast monsoon becomes active in the months of October, November and December which is the main rainfall season and so the main agricultural season of the region. This monsoon is responsible for about 55 per cent of total annual rainfall of the region (Table 3.1). Lower Palanis experiences heavy rainfall and the rainfall decreases towards the central stretch of the basin. In the Lower Palanis rainfall increases with altitude. In Berijam (2152 m MSL) the rainfall is 1437 mm and in Kodaikanal (2248 m MSL) it is 1615 mm. The Varushanad hills in the south also receive more than 2000 mm of rainfall while in the lowlands occupying the centre of the basin; rainfall is as low as 800 mm. The rainfall is minimum in the southwestern part of the lowlands around Rasingapuram (267 mm).

The summer and winter months are not of any significance with regard to rainfall, as these months together account for less than 10 per cent of the total annual rainfall. As such, they are generally dry months of the year in the basin.

3.2.1.4 Precipitation Ratio

The abnormalities of rainfall at any location may be brought out by a simple ratio of precipitation. It is the difference between maximum and minimum rainfall expressed in terms of the mean.* This ratio may give the stability level of rainfall. Higher the ratio, higher is the abnormality in rainfall and vice versa. The high abnormalities are found in the northwestern and southern part of the basin (Fig.3.6A and Table 3.2) (Bodinayakkanur, 270 per cent; Rasingapuram, 198 per cent; and Gudalur, 163 per cent). The precipitation ratio increases from the centre of the basin towards northeastern side. The lower abnormalities are recorded in other areas such as Kodaikanal (87 per cent) and Maruthanadi Dam (32 per cent).

3.2.1.5 Variability of Rainfall

Variability of rainfall about mean is worked out for annual and seasonal rainfall and is presented in Figures 3.6B and 3.7 and Table 3.2.

Fig. 3.6

* Precipitation Ratio = $\frac{Px - Pn}{Pm}$ x 100 where Px and Pn represent the maximum and minimum rainfall amounts over a series of years and Pm is the mean annual rainfall.

Variability is greater at the Thevaram and Lower Manjalar basins. In the former the rainfall variability is as high as 60 per cent while in the latter it is 50 per cent. These are the areas of low rainfall where mean annual rainfall is less than 700 mm. In general, variability increases westward. The north and northeastern parts of the Lower Palanis which receive the maximum rainfall has a variability of less than 20 per cent. The entire Thalai Vaigai, Andipatti, Viruveedu and Varahanadi basins have between 20 and 30 per cent rainfall variability. As generally believed areas of high rainfall experience low rainfall variability and areas of low rainfall have high rainfall variability.

3.2.1.5.2 *Seasonal Variability*

Figure 3.7 depicts the seasonal coefficients of rainfall variability in the basin. The northeast monsoon season, which is the main period of rainfall in this basin, records a rainfall variability ranging from 25 to 90 per cent. During the southwest monsoon period the variability varies from 23 to 151 per cent. A very high variability of rainfall is experienced during the dry months of winter when the variability ranges from 99-247 per cent, the average being 152 per cent. The maximum variability of rainfall is recorded at Rasingapuram in the Thevaram sub-basin.

3.2.1.6 Frequency of Rainfall amount

The frequency of rainfall amount of a station has been compared with the mean annual rainfall of the corresponding station. For twenty out of twenty four stations in and around the study area the particular class interval containing the normal rainfall has the highest frequency (Fig. 3.8). Hence the histograms take roughly the form of a pyramid in these stations. Thus excepting four stations (Rasingapuram, Berijam, Devadanapatti in the study area and Devikolam in Kerala) all other stations seem to be getting rainfall very close to the normal. The rainfall pyramids of the exceptional stations - Rasingapuram, Berijam, Devadanapatti and Devikolam - have two peaks on both sides of the normal class. Here, it is to be noted that Devikolam and Berijam and located on hills and Rasingapuram and Devadanapatti in the plains.

Table 3.2
Upper Vaigai Basin: Precipitation Ratio and Coefficient of Variability of Rainfall

Seasonal Variability of Rainfall						
Rain Gauge Station	Annual precipitation ratio	Winter	Summer	Southwest monsoon	Northeast monsoon	Annual Variability
Gudalur	162.85	191.37	61.49	57.10	43.77	35.81
Kambam	98.44	198.07	33.40	35.71	54.56	30.74
Uthampalaya,	100.88	119.21	50.99	56.74	33.36	25.80
Veerapandi	163.922	149.83	63.56	61.11	36.90	34.61
Rasingapuram	197.51	246.99	72.77	151.08	88.72	68.30
Bodinayakkanur	269.63	99.60	41.39	133.29	36.39	34.73
Berijam	175.39	128.80	63.42	37.86	55.13	35.34
Kodaikanal	87.36	98.53	38.44	25.45	34.26	17.20
Periyakulam	128.77	100.24	49.70	46.83	35.42	23.81
Kamatchipuram	89.28	191.43	52.11	48.19	48.50	30.05
Vaigai Dam	115.62	139.89	51.16	58.84	41.30	26.33
Manjalar Dam	123.84	169.79	57.23	60.29	49.54	30.86
Devadanapatti	141.98	200.35	47.11	76.76	65.53	42.74
Maruthanadi Dam	32.22	200.00	19.47	23.01	25.08	13.93
Peranai	84.1	132.28	54.00	45.79	35.19	21.41
Viralipatti	129.87	194.15	40.62	52.24	52.93	31.45
Kumili	113.13	121.62	53.02	36.64	41.58	26.81
Vandamettu	91.94	125.54	49.35	26.01	39.39	22.60
Chinnar	97.56	143.25	56.31	29.67	44.94	25.77
Devikolam	106.01	157.16	44.24	33.43	49.84	26.64
Nilakottai	97.85	148.71	52.48	49.58	32.10	22.30
Usilampatti	88.44	147.31	41.01	53.44	33.23	25.22
Watrap	91.37	118.86	54.26	94.24	29.73	25.45
Srivilliputhur	116.27	126.72	37.83	55.55	29.87	22.21
Upper Vaigai Basin	121.01	152.07	48.87	56.20	43.22	29.17

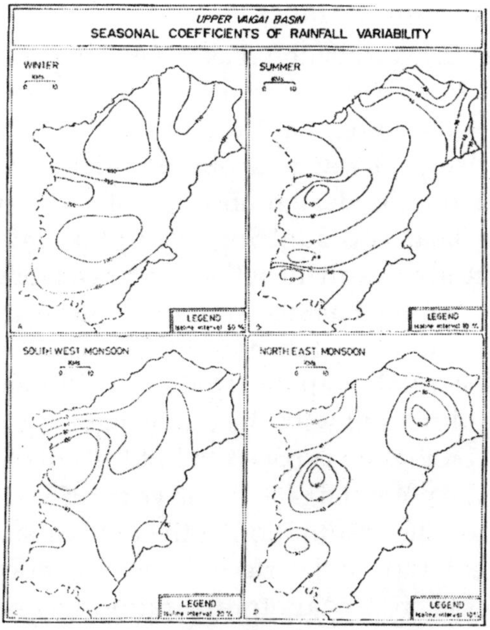
UPPER VAIGAI BASIN
SEASONAL COEFFICIENTS OF RAINFALL VARIABILITY
WINTER
SUMMER
SOUTH WEST MONSOON
NORTH EAST MONSOON
LEGEND
LEGEND
LEGEND
LEGEND

Fig. 3.7

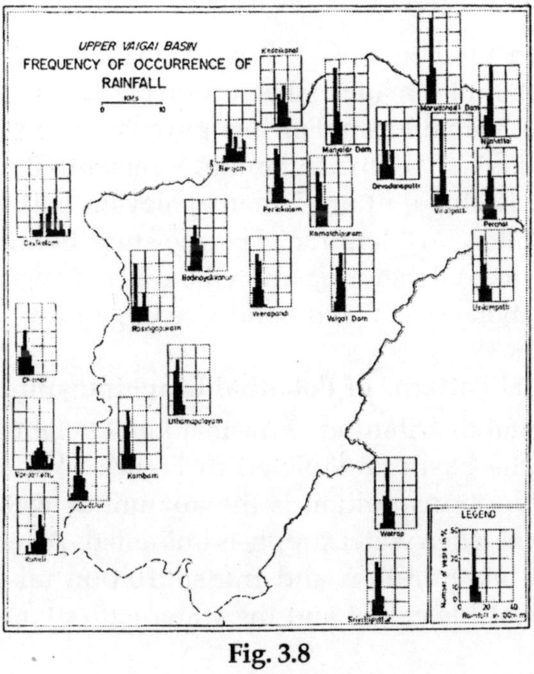
UPPER VAIGAI BASIN
FREQUENCY OF OCCURRENCE OF RAINFALL
KMs
Manjalar Dam
Periakulam
Kamatchipuram
Bodinayakanur
Vaigai Dam
Usilampatti
Uthamapalayam
Kombam
Watrap
LEGEND

Fig. 3.8

3.2.2 Climatic Water Balance

Water surplus is considered as the water potential of a region. It is generally held in the underground formations. It is partly responsible for the perennial flow of water in a river system. A detailed knowledge of water surplus and deficiency thus becomes a pre-requisite for the design and operation of the hydrological system. Water balance (budgeting of surplus and deficit) helps to have a meaningful assessment of the water availability of any region. The concept of water balance was introduced by Thornthwaite (1948). In this budgeting procedure the rainfall is treated on the income and potential evapotranspiration (PE) as the committed expenditure and rainfall with the due weight age of soil moisture. Soil is assumed to have a capacity to hold 10 cm of moisture which is available at the potential rate. In the year 1955, Thornthwaite and Mather modified the assumption of the field capacity of the soil so as to suite the soil type and surface cover. The ratio of utilization of soil moisture for crop evapotranspiration has been taken to be dependent on the actual soil moisture content rather considering it as available at the potential rate as was assumed in 1948.

Based on the book-keeping procedure of Thornthwaite and Mather an attempt is made in this section to study the water balance of the Upper Vaigai basin. For this purpose nine stations such as Kodaikanal, Berijam, Kamatchipuram, Viralipatti, Bodinayakkanur, Veerapandi, Rasingapuram, Uthamapalayam and Gudalur, have been taken into consideration. The moisture holding capacity of the soils are taken depending upon the soil and vegetation types of the stations (Soil Survey and Land Use Organization, 1980; 1981).

3.2.2.1 Spatial Patterns of Potential Evapotranspiration

The spatial distribution of mean annual potential evapotranspiration of the basin is depicted in Figure 3.9A and Table 3.3. Potential evapotranspiration is the amount of water evaporated and transpired when water supply is unlimited. When water supply is unlimited evaporation and transportation take place to the maximum possible extent and the water left out as balance is the surplus.

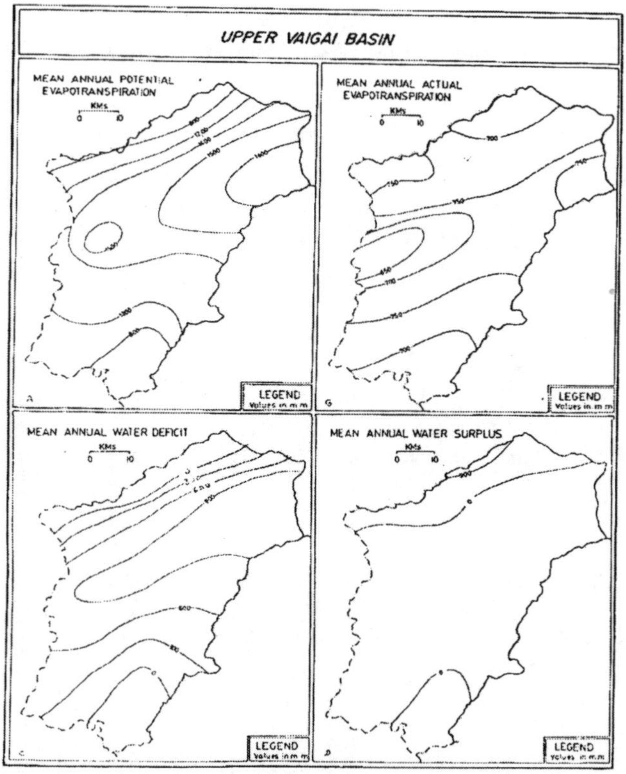

Fig. 3.9

The potential evapotranspiration is maximum (>1600 mm) in the central stretch and this is quite obvious as this stretch, covering about one third of the total area of the basin, supports a large concentration of population and their economic activities. On the other hand in the hilly areas of the Lower Palanis and the Varushanad the potential evapotranspiration or water need is 1200 mm and 800 mm respectively. In general, the potential evapotranspiration gradually decreases towards northwestern and southeastern sections of the Upper Vaigai basin. In the central stretch of the basin rainfall is limited to about 700 mm per annum and all this fall gets evaporated and transpired, thus leading to no surplus at all. The mean potential evapotranspiration of the basin is 1286 mm. This clearly indicates that this part of the basin faces a very serous problem of water deficiency.

Table 3.3
Upper Vaigai Basin; Water Balance

	Jan	Feb	Mar	Apr	May	June	July	Aug	Sept	Oct	Nov	Dec	Annual
Kodaikanal													
P	61	33	52	137	147	99	120	158	183	256	238	132	1616
PE	48	46	59	68	75	67	62	61	58	55	50	48	697
AE	48	46	59	68	75	67	62	61	58	55	50	48	697
WD	0	0	0	0	0	0	0	0	0	0	0	0	0
WS	13	0	0	49	72	32	58	97	125		188	84	919
Berijam										201			
P	37	29	53	134	102	61	112	137	167	289	197	124	1442
PE	48	46	65	68	75	67	62	64	61	49	50	48	703
AE	48	45	64	68	75	67	62	64	61		50	48	701
WD	0	1	1	0	0	0	0	0	0		0	0	2
WS	0	0	0	28	27	0	44	73	106	49	147	76	741
										0			
Kamatchipuarm										240			
P	5	16	37	44	66	21	19	34	122	168	141	66	739
PE	84	85	142	155	165	149	159	151	141		109	89	1554
AE	32	30	49	50	69	23	19	34	122	125	109	77	739
WD	52	55	93	105	96	126	137	117	19	12	0	12	812
WS	0	0	0	0	0	0	0	0	0	5	0	0	0
										0			
Viralipatti										0			
P	1	12	14	49	95	12	34	32	142	169	155	41	756
PE	80	123	152	174	173	161	152	147	142	145	123	72	1644
AE	20	26	22	52	96	12	34	32	142	145	123	52	756
WD	60	97	130	122	77	149	118	115	0	0	0	20	888
WS	0		0	0	0	0	0	0	0	0	0	0	0
Bodinayakkanur		0											
P	22	23	46	83	62	23	33	34	76	192	151	48	793
PE	78	85	127	139	152	143	149	148	141	132	97	92	1483
AE	45	42	63	91	72	32	37	36	77	132	97	71	795
WD	33	43	64	48	80	111	112	112	64	0	0	21	688
WS	0	0	0	0	0	0	0	0	0	0	0	0	0
Veerapandi													
P	16	19	40	59	44	16	21	23	142	148	95	69	692
PE	84	82	117	148	159	146	152	148	138	129	103	89	1495
AE	22	23	44	61	46	17	22	23	138	129	96	71	692
WD	62	59	73	87	113	129	130	125	0	0	7	18	803
WS	0	0	0	0	0	0	0	0	0	0	0	0	0
Rasingapuram													
P	5	2	9	38	54	66	35	46	95	90	160	17	617
PE	78	85	130	145	165	137	149	144	141	138	106	89	1507
AE	17	11	17	41	57	67	36	46	95	90	106	34	617
WD	61	74	113	104	108	70	113	98	46	48	0	55	890
WS	0	0	0	0	0	0	0	0	0	0	0	0	0
Uthamapalayam													
P	21	12	25	64	57	34	49	38	52	159	146	53	710
PE	87	82	108	121	136	127	136	125	119	116	97	95	1349
AE	34	28	39	70	63	39	52	39	53	116	97	80	710
WD	53	54	69	51	73	88	84	86	66	0	0	15	639
WS	0	0	0	0	0	0	0	0	0	0	0	0	0
Gudalur													
P	8	16	32	79	79	54	77	55	59	133	129	42	763
PE	72	79	99	105	123	102	107	103	101	95	85	74	1145
AE	30	31	44	83	84	58	79	58	61	95	85	55	763
WD	42	48	55	22	39	44	28	45	40	0	0	19	382
WS	0	0	0	0	0	0	0	0	0	0	0	0	0

P – Precipitation; PE – Potential Evapotranspiration; AE – Actual Evapotranspiration; WD – Water Deficiency WS – Water Surplus

3.2.2.2 Spatial Pattern of Actual Evapotranspiration

The actual evapotranspiration of water that is actually available for crop use depends on precipitation and soil moisture availability. When there is adequate amount of water it equals the potential evapotranspiration; when water is not adequate the actual evapotranspiration will be less than the potential evapotranspiration. The mean annual actual evapotranspiration is presented in Figure 3.9B and Table 3.3. The central stretch of the basin had recorded a very low actual evapotranspiration, which is equal to 700 mm of rainfall. Towards the foot hills and the highlands as the rainfall increases and the gap between actual evapotranspiration and potential evapotranspiration decreases. Thus there seems to be progressive surplus towards the foot hills and the highlands.

3.2.2.3 Water Balance

Lower Palanis and the Varushanad hills are virtually important in providing the Upper Vaigai basin with enormous amount of water surplus mainly on account of heavy rainfall in this region coupled with low water need (Figure 3.9D). The water deficiency is quite acute in the central stretch of the basin (Fig. 3.9C). It exceeds 850 mm in this region and ranges between 700 and 800 mm in the surrounding plain areas. The water deficiency is very low in the south and southeastern part of the basin. In the Lower Palanis water deficiency is negligible.

The elements of water balance of a station (AE, PE and P) when plotted graphically against the months of the year give a comprehensive picture not only of water surplus and deficiency but also of soil moisture accretion and depletion through the years. Figure 3.10 shows the located water balance graphs of different stations of the basin. Climatically Kodaikanal and Berijam have a surplus of 919 and 744 mm accumulated from April to December. The soil moisture is adequate to meet the water need fully in February and March when rainfall is below the potential evapotranspiration. All other stations have no surplus water during the months from January to August or September. The soil moisture is replenished during October and November to be utilized in other months; but this replenishment is not sufficient to avoid water deficiency. The mean cropping season of the basin is from August to December.

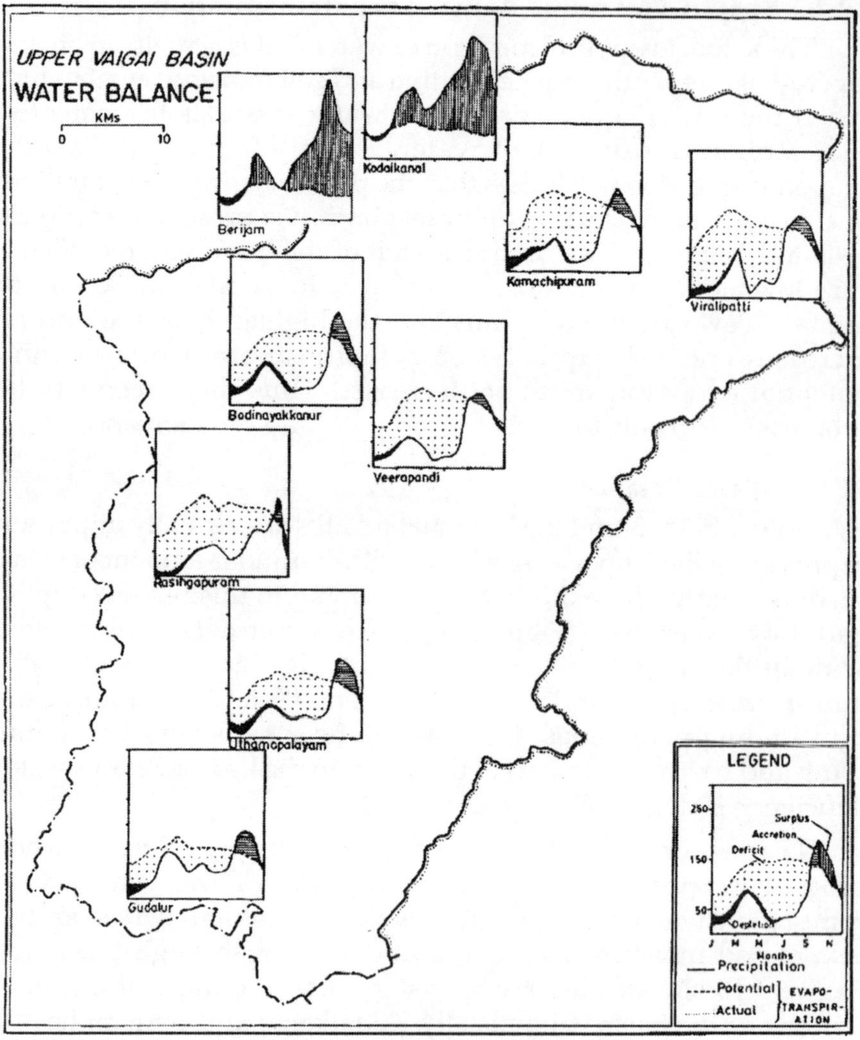

Fig. 3.10

Summing up it is apparent that water need is maximum in the central basin and decreases towards the northwestern and southeastern parts along the rim of hills of the basin. As such, it is essential to conserve water of the highlands and the foot hill zones in order to make it available to the thirsty lands of the central part of the basin.

3.2.3 River Flow Characteristics

One of the important aims of hydrology is to analyze the spatial and temporal variations in river flow characteristics, particularly when a basin hydrology is dominated by random rainfall, as in the monsoon regime. The total annual river flow varies from year to year and knowledge of variability of river flow is essential for any water resource studies. Monthly river flow data are collected from four stations along the river Vaigai. The selected stations are Power House, Vaigai Dam, Manjalar Dam and Peranai (Fig. 3.1)*.

3.2.3.1 Determination of River Flows at Various Percentage of Dependabilities

The design criteria for any water resource development project are normally fixed for a specified dependability of annual yield in the river. For example, an irrigation project is normally designed based on a 75 per cent dependable annual yield which would mean that for a successful irrigation the annual river flow must be available for three out of four years on an average.

The percentage dependabilities are worked out by statistical analysis of a runoff series and tested to see whether it follows a normal or a skewed distribution

3.2.3.1.1 Frequency Distribution of Annual River Flows

3.2.3.1.1.1 Log Normal Distribution

The log normal distribution function is widely used in the analysis of frequency distribution of river flows (De Ridder and Erez, 1977). The basic justification for this distribution function is the central limit theorem, which states that the logarithm of a variable which is the sum of identically distributed random variables derived from any distribution with a finite mean and variance is distributed normally. The annual river flows at Power House, Vaigai Dam and

---

* Time-series are not available uniformly for all these stations.

	Station	Years of Data
1.	Power House	1948 - 2002
2.	Vaigai Dam	1900 - 2002
3.	Peranai	1948 - 2002
4.	Manjalar Dam	1968 - 2002

Peranai* have been plotted on a probability paper of which the probability axis is constructed in such a way that the cumulative normal distribution

$$F(P < X) = 85X \; f(x)\; dx$$

is presented as a straight line (Fig.3.11 A, B, and C) The plotting positions are obtained by using Tables in Appendices 3.3, 3.4 and 3.5. In order to obtain a straight line the parameters are estimated as m = P and s = sp from the annual river flows according to

$$EST(\mu) = \bar{P} = \frac{\Sigma P}{n} \qquad EST(\sigma) = SP = \frac{\sqrt{P2 - n(\bar{P})2}}{n - 1}$$

When the values of P = μ - s of the three different stations are then plotted against the 16 per cent and 84 per cent non-exceedence probability whereas the value of P = μ is plotted against the 50 per cent probability. A straight line is drawn through these positions.

Figure 3.11 A, B, and C shows the plots of three river flow stations at Power House, Vaigai Dam and Peranai. It also shows that the lines of the theoretical log normal distribution, calculated with mean and variance. These lines match the points reasonably well. In the first instance it is interesting in that the total annual stream flows have a return period of 20, 5, and 1.67 years corresponding with an exceedance frequency of 95, 80 and 40 per cent respectively. The frequencies referred to rivers flows at Power House, Vaigai Dam and Peranai are present in Table 3.4.

The analysis has demonstrated that the mean annual river flow has recurrence intervals of 20, 5 and 1.67 years i.e. on an average, the annual maximum flows will exceed once in every 20, 5 and 1.67 years respectively and this recurrence intervals are indicated by broken lines in Figure 3.11 A, B and C.

3.2.3.1.1.2 *Gumbel Distribution*

For reasons of comparison, it is also assumed that the total river flow at Power House, Vaigai Dam and Peranai have Gumbel

* Frequency distribution of annual river flow at Manjalar Dam is not considered because of the limited availability of data.

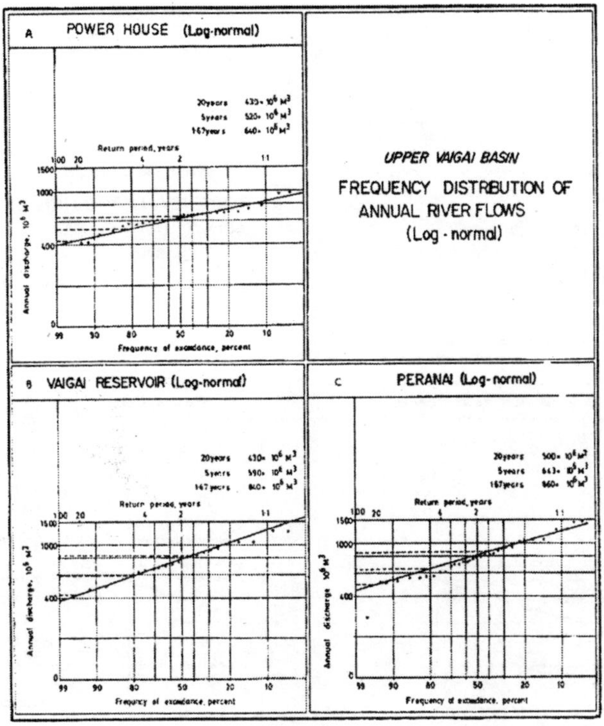

Fig. 3.11

Table 3.4

River Flows For Three Return Periods (By Log Normal Distribution)

River Flow Station	River Flows for three Return periods 20	5 (in million m^3)	1.67
Power House	430	520	640
Vaigai Dam	430	590	840
Peranai	530	643	860

distribution, which is a special case of log normal distribution (Gumbel, 1954). The cumulative frequencies are plotted on the Gumbel probability papers in Figure 3.12 A, B and C. For the estimation of straight line the following equation is used,

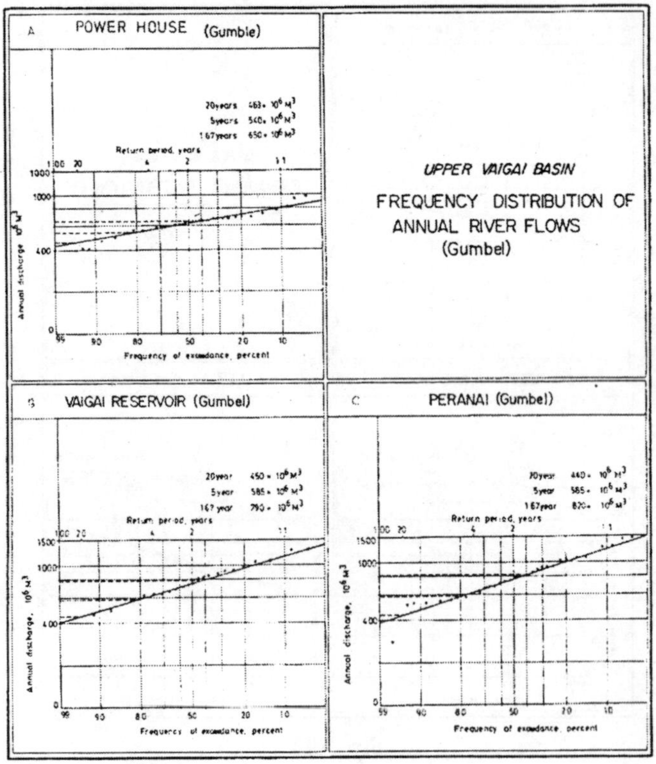

Figure 3.12

$$\text{EST}(\alpha) = \frac{\text{Sn}}{\text{Sp}} \quad \text{and} \quad \text{EST}(\mu) = \bar{\text{p}}\text{n} - \frac{\bar{\text{y}}\text{n}}{\text{ESL}(\alpha)}$$

The frequencies referred to river flow at Power House, Vaigai Dam and Peranai are presented in Table 3.5.

3.2.3.2 Monthly River Flow: Flow Duration Curves, Central Tendency and Deviation

The monthly river flow data for four stations namely Power House, Vaigai Dam, Manjalar and Peranai have been analyzed by constructing flow duration curves to study and compare the magnitudes of river flow. Further an attempt has also been made to study the mean, maximum, minimum standard deviation and coefficient of variability and monthly river flow at these stations.

Table 3.5
River Flows For Three Return Periods (By Gumbel Distribution)

River Flow Station	River Flows for Three Return periods 20	5	1.67 (in million m³)
Power House	463	540	650
Vaigai Dam	450	585	790
Peranai	440	585	820

3.2.3.2.1 *Flow Duration Curves*

The flow characteristics of a river through the range of discharge independent of chronological sequence may be studied by means of flow duration curves. The river flow data are arranged in order of magnitude and plotted on arithmetic scale as per cent of time during which they were equaled or exceeded. This slope is also a measure of river flow variability (Searcy, 1959). Flow available at 90 per cent of time (q 90) and flow available at 50 per cent of time (q 90) are the measures of flow variability.

The flow duration curves of mean monthly flows at Power House, Vaigai Dam, Manjalar Dam and Peranai are presented in Figure 3.13 A, B, C and D respectively. A relatively large decline in the frequency of recurrence of river flows between 500 and 120 million m^3 in Peranai, 392 and 140 million m^3 in Vaigai Dam and 9.4 and 4 million m^3 in Manjalar Dam are noticed whereas in the Power House the decline is negligible comparatively i.e. between 140 and 120 million m^3.

A large decline in the magnitudes of higher river flows is followed by only a marginal increase in the magnitudes of lower flows, particularly below the range of 120 million m^3 in the Power House and Peranai and 140 million m^3 in the Vaigai Dam and 4 million m^3 in the Majalar Dam (Table 3.6). The frequency of stream flows available unto 62 per cent of time has been slightly increased.

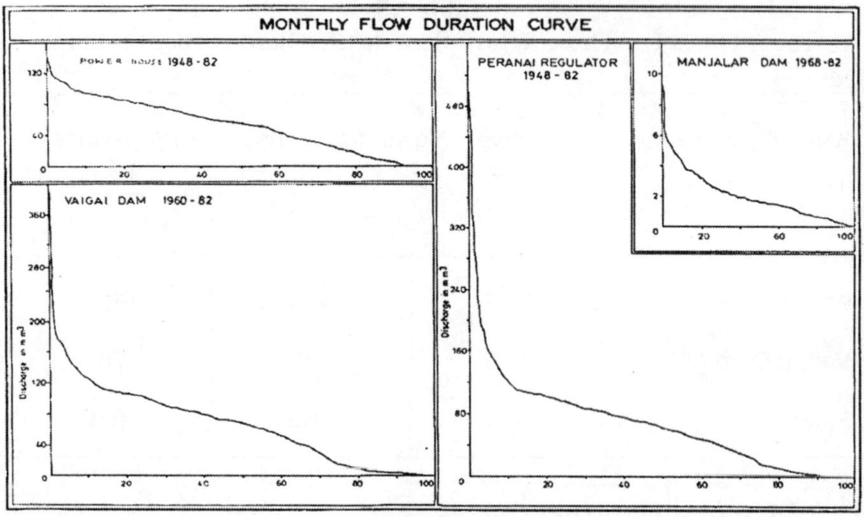

Fig. 3.13

Table 3.6
River Flows Available at 50 and 90 per cent of Time

River Flow Station	River Flows Available at 90 per cent of time	at 50 per cent of time
	(in million m³)	
Power House	6.00	66.00
Vaigai Dam	4.00	68.00
Manjalar	0.30	1.60
Peranai	2.00	62.00

Variability indices based on Lane's method have been computed for these four flow duration curves (Searcy, 1959). The index is given by standard deviation of the logarithm at 5 per cent to 95 per cent river flows with an interval of 10 per cent between the successive values. However, the variability index determined from a flow duration curve varies with a length of record owing to the effect of persistence in hydrologic data (Chow, 1964). The computed variability indices for Power House, Vaigai Dam, Manjalar and Peranai are 0.652, 0.619, 0.405 and 0.684 respectively. The computed variability indices for these three regulated flows suggest very little effort on the part of Power House, Vaigai Dam, Manjalar and Peranai in controlling variability of the mean monthly flows even after operation of the reservoirs and flow regulations through canals.

3.2.3.2.2 *Monthly River Flow: Central Tendency and Deviations*

The monthly maximum, minimum and mean and variability of river flows are presented in Table 3.7 for the four stations Power House, Vaigai Dam, Manjalar and Peranai. The Figure 3.14A presents the mean river flows. The river flow is very low in the months of March, April and May the practically dry months of the year with no rainfall and starts increasing from May onwards. It reaches its peak during the northeast monsoon in the month of November in Peranai (149.96 million m^3) and Vaigai Dam (123.57 million m^3) and two months earlier in September in the Power House (83.17 million m^3). After this peak, the flow decreases. The flow is very low between April and August in Manjalar (Fig.3.14C) with the least in August (0.74 million m^3) and from then on it increases to a maximum in November (4.75 million m^3) with the arrival of the Northeast monsoon.

The maximum annual for Power House, Vaigai Dam, Manjalar and Peranai are 962.65, 1263.20, 40.00 and 1457.90 million m^3 respectively. The minimum annual discharge for these stations is 222.90, 261.35, 15.82 and 278.24 million m^3 respectively (Table 3.7). The monthly coefficient of variability of river flow for these stations is very high in the month of May (Power House, 152.37; Vaigai Dam, 132.96; and Peranai 254.71 per cent). From January the variability increases and in the month of March it decreases in two stations namely Peranai and Vaigai Dam. It reaches its peak in May. From May the variability decreases gradually. The variability is quite large form September to November. Owing to the precarious rainfall during the summer season the maximum variability occurs between March and May (Fig.3.14B) and then onwards it decreases

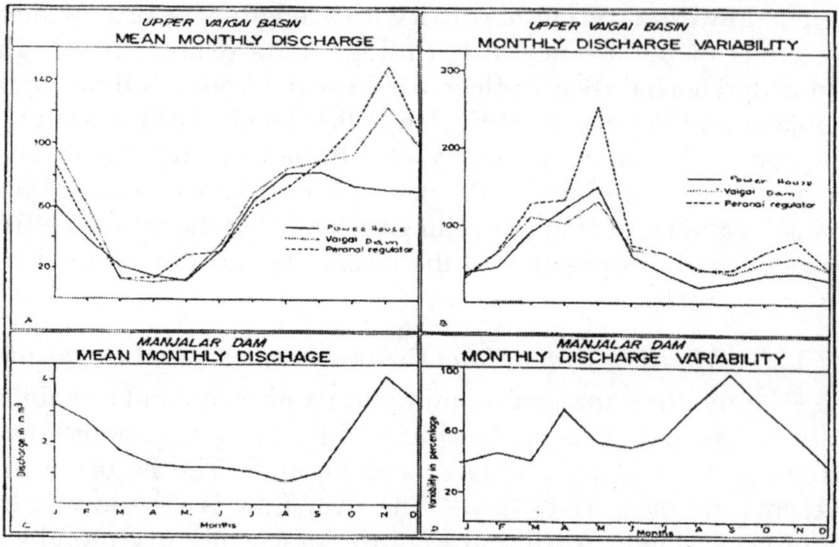

Fig. 3.14

to a maximum for June to September in all the stations. This is because of the regulated supply of water during the crop season.

Figure 3.14D shows variability of the Manjalar river flow. The variability is more in the months of May and from August to November. From December to February, variability is very low because of regulated supply of water during these months of crop season.

Summing up, during the summer season the discharge is very low and the coefficient of variability is very high. But from June to November the amount of river flows at Power House, Vaigai Dam and Peranai increases gradually but the variability is decreasing due to the very heavy rainfall in the catchments areas (Kumili, 1203 mm and Vandamettu, 1657 mm rainfall) during this period, thus indicating a greater stability of flows in these months. During the northeast monsoon season, the variability is slightly high. Inverse is the case in Manjalar and it may be because of heavy rain during the northeast monsoon (Manjalar Dam: mean northeast monsoon seasonal rainfall is 337 m mm which contribute 46 per cent of annual rainfall).

Table 3.7

Upper Vaigai Basin; River Flow: Central Tendency and Deviations

River Flow Gauge	Jan	Feb	Mar	Apr	May	June	July	Aug	Sept	Oct	Nov	Dec	Annual
PERIYAR DAM													
Mean	60.52	39.41	20.69	14.77	12.47	30.86	65.00	82.28	83.17	74.33	72.24	71.12	626.86
Maximum River Flow	105.56	70.40	66.09	68.84	68.84	79.06	131.81	108.51	112.24	119.0	118.36	107.77	962.65
Minimum River Flow	16.87	7.79	0	0	0	0	9.15	33.07	21.18	12.03	15.97	21.46	222.90
Standard Deviation	22.60	17.86	18.80	17.63	18.73	18.45	24.17	16.61	20.80	26.70	26.97	18.93	132.95
Coefficient of variability in per cent	37.89	45.98	92.22	121.12	152.37	60.63	37.74	20.48	25.38	36.45	37.87	27.01	21.52
VAIGAI RESERVIOR													
Mean	97.65	60.85	12.12	10.51	13.26	36.19	70.08	84.82	90.40	93.29	123.57	98.10	790.84
Maximum River Flow	163.10	121.93	55.42	35.11	59.75	72.49	214.07	163.67	156.59	139.6	273.97	177.66	1263.0
Minimum River Flow	14.98	2.58	0	0	0	1.36	0.93	1.39	2.29	0	70.28	42.81	261.35
Standard Deviation	28.77	37.76	13.46	10.80	17.24	23.97	44.52	3.10	33.82	47.16	70.95	37.46	240.16
Coefficient of variability in per cent	30.13	63.45	113.61	105.04	132.96	67.70	64.95	43.54	38.25	51.69	58.70	39.04	31.05
MANJALAR DAM													
Mean	3.24	2.68	1.77	1.29	1.19	1.18	1.03	0.74	1.03	2.57	4.15	3.27	24.14
Maximum River Flow	5.47	5.35	2.83	3.79	2.41	1.95	1.93	1.90	3.51	7.96	9.20	5.58	40.00
	1.53	1.19	0.42	0	0.08	0.03	0.06	0.11	0	0	0.79	1.50	15.82

Minimum River Flow													
Standard Deviation	1.23	1.20	0.70	0.93	0.61	0.57	0.55	0.56	0.95	1.88	2.28	1.14	6.90
Coefficient of variability in per cent	39.51	46.27	41.24	74.42	52.94	50.00	55.34	78.38	96.11	75.49	56.87	36.09	29.56
PERANAI REGULATOR													
Mean	85.92	48.35	12.24	14.13	29.17	30.75	61.38	73.48	98.54	112.4	149.96	105.96	813.56
Maximum River Flow	126.09	118.19	77.90	68.98	421.92	76.85	210.67	148.66	227.47	446.8	502.36	190.40	1457.0
Minimum River Flow	11.19	1.76	0	0	0	0	0.33	0.59	2.35	3.20	40.24	47.63	278.24
Standard Deviation	29.54	31.29	16.22	18.79	73.24	22.63	39.17	29.76	37.31	72.69	118.67	42.10	266.03
Coefficient of variability in per cent	34.89	65.67	129.73	133.89	254.71	74.67	64.74	41.10	42.27	65.70	80.29	40.31	33.18

3.2.4 Upper Vaigai Sub-basins: Surface Water Resources

The river flow analysis, so far discussed in the preceeding sections pertains to the Upper Vaigai basin, as a whole. The flow is the combined contribution of sub-basin to the major basin. The basin has eleven sub-basins and extending the flow analysis to sub-basins may give a further understanding in the flow characteristics of the basin.

As an actual rainfall fluctuates in time and space a spatial rainfall distribution in many cases cannot always be considered uniform. Since there are many techniques of runoff prediction and forecasting some of these are in widespread use, either because they work reasonably well over a wide range of conditions or else are easy to apply. The use of other techniques may be restricted to specific areas or to specific users, such as a particular government agency. In order to obtain the sub-basin-wise quantity of surface water an attempt has been made to study the runoff of the entire Upper Vaigai basin based on the empirical method namely Thiessen polygon method.

The Thiessen polygon method was first devised by Thiessen and later perfected by Horton (Thiessen, 1924). It is used to determine the mean depth of precipitation over a given area. The use of Thiessen polygon makes some allowance for the uneven distribution of gauges throughout the area, and also enables data from adjacent areas to be incorporated in the mean. In this method perpendicular bisectors are drawn through the straight lines joining adjacent gauges, leaving each gauge in the centre of a polygon which will vary in size according to the spacing of gauges (Fig.3.15).

The yield rainfall for each station is derived at from the Stranger's table for the corresponding weighted mean monsoon precipitation (Strange, 1928). The yield is calculated after allowing one third of the weighted average rainfall as evaporation loss (Public Works Department, 1981). The total mean annual yield of surface water in the Upper Vaigai basin is 313.952 million m^3 (Table 3.8). Very high yield is found in the Varahanadi basin (68.692 million m^3) followed by the Thalai Vaigai basin (57.326 million m^3), the Suruliar (52.494 million m^3) and the Kottagudi basin (49.338 million m^3). The surface water yield is low in the Viralipatti basin (11.211 million m^3)

Fig. 3.15

3.2.5 Surface Water Quality

It is tried to understand chemical composition of the surface water and its relevance to different purposes in the basin. The surface water may broadly be classified into cation and anion groups. The cation group is further identified as calcium type, magnesium type, sodium type and intermediate type depending upon the predominance of the chemical constituents and anion group as bicarbonate type, sulphate type, chloride type and intermediate type (Table 3.9).

Table 3.8

Upper Vaigai Basin; Perceage of Runoff, Depth and Yield of Runoff

Name of the sub-basin	Area in km^2	Weighted mean rainfall in mm	Rainfall after evaporation loss 2/3 of weighted mean	Percentage of runoff	Depth	Yield of runoff from basin per km^2 in m^3	Mean Annual yield in million m^3
Thalai Vaigai	842.16	707.35	471.57	11.91	57.84	57795.68	57.326
Suruliar	696.16	725.86	483.90	11.92	59.88	59826.20	52.494
Thevaram	261.87	652.91	435.27	8.83	36.46	38403.10	10.203
Kottagudi	402.71	882.58	588.39	16.84	106.10	106011.08	49.338
Koilpatti	136.29	704.61	409.74	9.99	47.02	47006.30	6.406
Varahanadi	419.52	980.89	653.92	19.54	138.98	138882.25	68.692
Manjalar	264.69	799.69	533.12	14.13	75.72	75660.05	19.969
Maruthanadi	244.40	879.82	386.55	16.28	96.42	96333.02	22.819
Viralipatti	66.74	758.08	505.39	11.17	56.46	56401.90	3.764
Viruveedu	148.03	774.06	516.04	15.35	79.30	79240.00	11.730
Andipatti	248.67	637.38	425.05	11.30	48.08	45081.90	11.211
Upper Vaigai Basin		773.02	497.18	13.39	81.06	72785.67	313.953

Surface water samples are collected from 12 sites distributed all along the river Vaigai and its tributaries (Fig.3.16). The major chemical constituents like Ca, Mg, Na, K, HC 03, C 03, S 04 and Cl are determined. Other tests like chemical conductivity (Ec) Hydrogen ion concentration (pH), total dissolved solids (TDS), and total hardness as $CaCO_3$ and sodium adsorption ratio (SAR) are also conducted to evaluate the quality of surface water of the study area. Normally observations taken in July are representative of pre-monsoon conditions which those taken in January represent post-monsoon conditions.

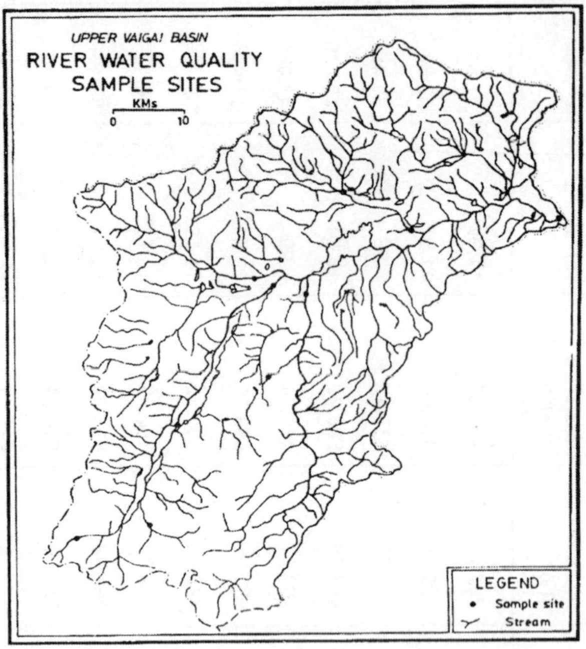

Fig. 3.16

3.2.5.1 Classification of Surface Waters

Among the cation group of surface waters calcium have been found in six sample sites in July and one in January. Sodium and potassium types are found in one sample in January but it is not found in July. Intermediate type is dominant among the samples in the surface water samples. Bicarbonate type is found in ten sites in January and eleven sites in July. Chloride is found in only one site in July.

Table 3.9
Upper Vaigai Basin: Surface Water Types

Surface Water Types	Number of Samples in 2002	
	January	July
CATION GROUP		
Calcium	1	6
Magnesium	-	-
Sodium & Potassium	1	-
Intermediate	10	6
ANION GROUP		
Carbonate	-	-
Bicarbonate	10	11
Chloride	-	1
Intermediate	2	-

3.2.5.2 Salinity and Alkalinity

3.2.5.2.1 Hydrogen Ion Concentration (pH)

The pH values range form 7.7 to 8.4 (Table 3.10), mean being 7.45, thus indicating a strong alkalinity of river water.*

3.2.5.2.2 Total Dissolved Solids (TDS)

The mean TDS of the surface water of the basin is 328.65 parts per million (ppm). The average TDS in sodium types 342.50 ppm, Calcium type 173 ppm, bicarbonate type 295.20 ppm and intermediate type 490.50 ppm.**

3.2.5.2.3 Electrical Conductivity (EC)

The total concentration of soluble salts in surface water can be measured in terms of electrical conductivity. Usually the upper limit taken for irrigation water is 5000 EC (Micro ohms/cm at 25° C) (Richards, 1954). The mean EC of surface waters in the basin is 514.10. The mean EC in calcium type of surface water is 270.00; in sodium and intermediate type in cation group, 535.00; and in bi-carbonate type 462.50; and intermediate type in anion group, 766.00.

---

* Neutrai water has a pH of 7.0; alkaline water has a pH of more than 7 and an acidic water less than 7 pH value (Walton, 1970).

** The total dissolved solids in a water sample include all solid materials in solution and from 10 ppm in rainwater to more than 3, 00,000 ppm in brines.

Table 3.10

Upper Vaigai Basin:. Major Chemical Characteristics of Surface Water

(Values in equivalent parts per million)

Variables	Calcium type*	Sodium Type			Intermediate Type		
		Mean	Max .	Min.	Mean	Max.	Min.
Calcium	30.00	23.00	28.00	18.00	36.67	54.00	24.00
Magnesium	12.20	24.50	38.90	10.09	24.73	45.00	10.90
Sodium	9.20	75.90	94.30	57.50	50.09	92.00	9.20
Carbonate	-	51.00	90.00	12.00	35.33	180.00	12.00
Bicarbonate	134.00	229.00	244.00	214.00	281.89	372.00	116.00
Chloride	24.80	37.25	42.60	31.90	49.27	92.20	21.30
PH	7.90	8.25	8.30	8.20	7.96	8.20	7.70
TDS	173.00	342.50	465.00	220.00	341.67	550.00	115.00
EC	272.00	535.00	725.00	345.00	535.00	862.00	181.00
SAR	0.36	2.60	2.90	2.30	1.49	2.80	0.75
Total Hardness as CaCo3	125.00	160.00	205.00	115.00	193.33	320.00	105.00
Na in per cent	13.79	31.04	52.08	50.00	32.59	49.38	16.00
RSC	-	2.25	2.90	1.60	1.46	6.70	0.10

* Only one sample analyzed

Anion Group

Variables	Bicarbonate Type			Intermediate Type		
	Mean	Max .	Min.	Mean	Max.	Min.
Calcium	36.20	54.00	24.00	22.00	26.00	18.00
Magnesium	21.08	45.00	10.09	36.50	38.90	34.10
Sodium	42.55	87.40	9.20	93.15	94.30	92.00
Carbonate	27.00	180.00	12.00	75.00	90.00	60.00
Bicarbonate	232.90	372.00	116.00	241.00	244.00	238.00
Chloride	42.56	92.20	21.30	58.55	74.50	42.60
PH	7.96	8.20	7.70	8.20	8.30	8.10
TDS	295.20	550.00	115.00	490.50	516.00	465.00
EC	462.50	862.00	181.00	766.00	807.00	725.00
SAR	1.33	2.60	0.36	2.85	2.90	2.80
Total Hardness as CaCo3	177.50	320.00	105.00	205.00	205.00	205.00
Na in per cent	30.98	52.00	13.79	49.69	50.00	49.38
RSC	1.78	6.70	0.10	2.40	2.90	1.90

3.2.5.3 Sodium Adsorption Ratio (SAR)

It is an index to measure sodium or alkaline hazard of the surface water used for irrigation.* Normally SAR ranges from 1 to 32. But in a few places of the basin the ratio exceeds 100. Very high mean SAR is found in intermediate type of anion group (285); followed by sodium type (260), intermediate cation type (1.49), bicarbonate type (1.33) and calcium type (0.36). The mean sodium adsorption ratio of the surface waters of the basin is 1.72.

3.2.5.4 Total Hardness

The total hardness of the surface waters may be evaluated from the concentration of $CaCO_3$ in the surface waters. The average hardness of surface waters is above 125.00 in calcium type; 160.00 in sodium type, 193.33 in intermediate (cation) type, 177.50 in bicarbonate type and 205.00 in intermediate anion type. The mean hardness in all the surface waters in the basin is 172.16.

From the above surface water quality analysis it is inferred that the surface waters of the Upper Vaigai basin are classified as good to permissible for irrigation, domestic and industrial supply.

3.3 GROUNDWATER RESOURCES

Groundwater resources refer to mean water occupying all the voids within a geologic stratum (Todd, 1959). It is replenish able and dynamic and is one of the Earth's most widely distributed resources. Although generally renewable, its availability is often limited with respect to time and space. These limitations may become all the more prominent on account of man's ever-increasing interference with the hydrological cycle (Walton, 1970).

The quantum of surface water has not been found to cope with the increased demand with the result that emphasis has now shifted to the exploitation of groundwater (Sharma, 1971). Shortages of even groundwater in areas of excessive drought emphasize the importance of correct estimates and proper development,

* The sodium adsorption ratio is calculated as

$SAR = Na/\sqrt{ca + Mg}\ /\ 2$

where Na, Ca and Mg represent the concentration in milli equivalents per liter.

regulation and protection of supplies in order to ensure the continued availability of this key natural resource (Todd, 1959).

3.3.1 Groundwater in the Hydrologic Cycle

Groundwater constitutes a part of the hydrologic cycle. Water bearing formations of the Earth's crust act as conduits for transmission and as reservoirs for storage of water. Groundwater emerging into surface stream channels aids in sustaining stream flow. Practically all groundwater originates as surface waters. Principal sources of natural recharge include precipitation, stream flow, lakes, reservoirs and canals. Other contribution known as artificial recharge occur from excess irrigation. Water enters this formation from the surface or from bodies of surface water, after traveling slowly for varying distances; it returns to the surface by the action of natural flow, plants or man.

Water within the ground moves downward through the unsaturated zone under the action of gravity, whereas in the saturated zone it moves in a direction determined by the surrounding hydraulic situation. Discharge of groundwater occurs when water emerges from underground. Most natural discharge occurs as springs into surface or surface water bodies, such as streams, lakes and oceans groundwater near the surface may return directly to the atmosphere by evaporation from within the soil and by transpiration from vegetation. Pumping from wells constitutes the major artificial discharge of groundwater.

An attempt has been made in this section to study the groundwater resources of the basin, based on the secondary data made available by agencies like Groundwater Division, Central Groundwater Board and Geological Survey of India.

3.3.2 Hydrologic Properties of Water Bearing Materials

The principal hydrologic properties of a rock formation are its porosity, specific yield and permeability which control the capacity of a rock to hold, deliver and transmit water. They exhibit assortment and concentration of the constituent grains. For example, a clear coarse sand or gravel possesses high porosity, specific yields and permeability. It absorbs water readily stores it in large quantities and yields it to wells freely whereas a poorly assorted sand hardly exhibits any of these qualities. However, the bulk of the material is not well assorted and consists of a mixture of sand and clay in varying proportions.

3.3.2.1 Seepage

The term seepage is defined as the flow of water through soils (Wesseling, 1972). While working for groundwater potential, it is important to study the seepage in the study area. The seepage varies from one place to another. Based on the studies conducted by the Groundwater Division in various places in the study area the value of seepage generally taken is 90 cm. The tank seepage studies conducted in Kelapoolandapuram, Veriyappanayakkanpatti and Sankarapuram give the value of tank seepage as 45 cm. (Public Works Department, 1976).

3.3.2.2 Porosity

The amount of groundwater stored in saturated material depends upon its porosity. The porosity of a rock or soil is defined as the property of containing interstices and is normally expressed in terms of the percentage of the total volume of materials which is represented by its interstices. Well assorted deposits of uncemented sands and gravels possess high porosities as against the poorly assorted materials. The porosities of the alluvial material are generally rather low on account of the unassorted nature of most of the depositions.

3.3.2.3 Specific Yield

This term was proposed for the first time by Meinger (1923) to designate the available water free, to drain out of a rock under natural conditions and it is expressed as percentage of the total volume of the sediments. All groundwater in the rock interstices is not, however, available for recovery. This is, in fact, limited by the specific yield of the rock formation. Specific yield is expressed quantitatively as the percentage of the total volume of sediments drained as indicated by a decline in water table, or the percentage of the total volume of sediments saturated by recharge as indicated by a rise in water table.

The unconsolidated and well-assorted coarse sands and gravels possess high specific yields as they readily part with a large proportion of their groundwater content. They are, therefore, largely tapped for obtaining water. On the other hand, fine materials for instance, clay or unassorted or poorly assorted materials although possessing comparatively high porosities, are poor in water yield due to the low specific yields. Thus the quantity of groundwater of a formation as a source of water supply will mainly depend upon its specific yield.

Table 3.11 presents the specific yield capacity in per cent in different formations of the study area. From this table, it is observed that in the areas of low specific yield capacities, the groundwater table is very deep and vise versa.

Table 3.11

Upper Vaigai Basin: Specific Yield of Different Formations

Formation	Test site	Specific yield in per cent	Mean Groundwater Table in m
Garnetiferous gneiss	Andipatti	4.38	6.87
Weathered biotite gneiss	Errasakkanayakkanur	2.45	14.75
Charnockite	Venkatachalapuram	3.67	11.81
Joint Charnockite	Kombaipatti	8.77	7.02
Fissured Charnockite	Viralipatti	2.87	8.23
Fissured Charnockite	Gullapuram	9.57	7.52
Fissured Granite	Vadapudupatti	1.38	15.24

In the central part of the study area, groundwater table is more than 16 m deep below the ground level due to the poor specific yield capacity (Venkatachalapuram, 3.67 per cent) of charnockite formations.

The groundwater depth is very low in the Kottagudi, Andipatti and southeastern part of the Upper Suruliar sub-basins. The specific yield capacity is the main controlling factor (5.6 per cent) in these areas.

3.3.2.4 Geophysical Investigations

The geophysical investigations are being increasingly employed in the analysis of water resources (Sharma, 1971) and the main aim of the geophysical survey is to demarcate suitable areas for groundwater development. The depth of the probing was restricted to 100 m. the depth soundings are selected at the interval of 3.2 km along the transverse lines, which are drawn with an interval of 3.2 km and perpendicular to the regional strike direction.

Based on the resistively survey at 10 and 30 m the low resistivity areas are identified. As per the resistivity survey the depth to bed rock varies from 5 to 50 m below ground level and the depth to the bed rock is more where the thickness of the alluvial top soil is greater as in the river beds and hill slopes. In general the depth to the bed rock ranges from 10 to 20 m below ground level in these places. The rock types encountered with in this area are granites and gneisses. A few veins of pegmatite's and quartzite cut across the country rock. Fractures and joints are more prominent in quartzite and granite.

Based on the geophysical investigation the low resistivity zones are identified (Fig.3.17). The groundwater potential is more in low resistivity zones. The Andipatti and parts of Seepalakottai micro basins have greater groundwater potential. The Upper Suruliar basin covering the villages of Hanumanthanpatti, Pudupatti, and Uthamapuram belong to the category of low resistivity zones. The western Thevaram basin covering the villages of Thevaram, Pannaipuram, T. Meenakshipuram and the western part of Kombai have low resistivity. The lower most part the Suruliar basin covering the villages of Dombuchery, B. Ammapatti and Uppukottai have also low resistivity and therefore good groundwater potential. In the Varahanadi basin, Vadakarai, Thenkarai, Thamaraikulam and Periyakulam villages have good groundwater potential since these areas are generally of the low resistivity type. The western part of the Manjalar and a few patches here and there in Maruthanadi basin also have good groundwater potential (Fig. 3.17). Owing to the hilly nature of the country the north and southeastern part of the study area have not been covered by the survey.

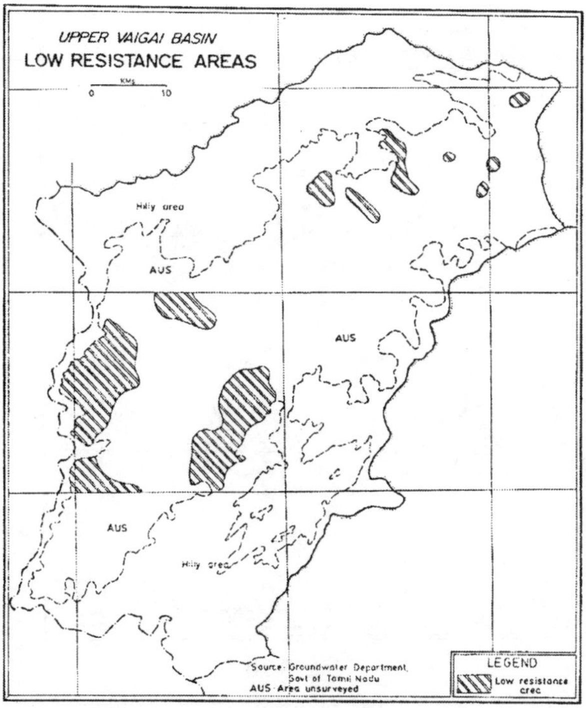

Fig. 3.17

Figure 3.18 and 3.19 depict the isoresistivity at 10 m and 30 m of depth respectively. The high resistivity areas are seen in both figures at the centre (more than 700 ohm metres in 10 m depth and 500 ohm metres in 30 m depth). Another zone located south in the Lower Suruliar basin and eastern part of the Thevaram basin have the value of 400 ohm metres in both Figures 3.18 and 3.19. The third zone may be noticed in the Lower Varahanadi basin which was a high resistivity of more than 400 ohm metres in 10 m depth and 600 ohm metres in 30 m depth isoresistivity maps. Needless to say that all these areas have a very low groundwater potential.

3.3.3 Groundwater Table and its Characteristics

Groundwater table is a physical surface defining the upper limit of the zone of saturation of water. The water table is an indicator of the hydraulic conditions prevailing in the zone of

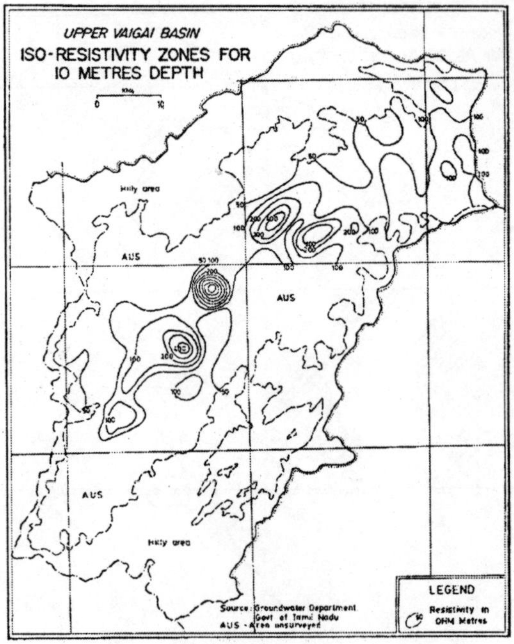

Fig. 3.18

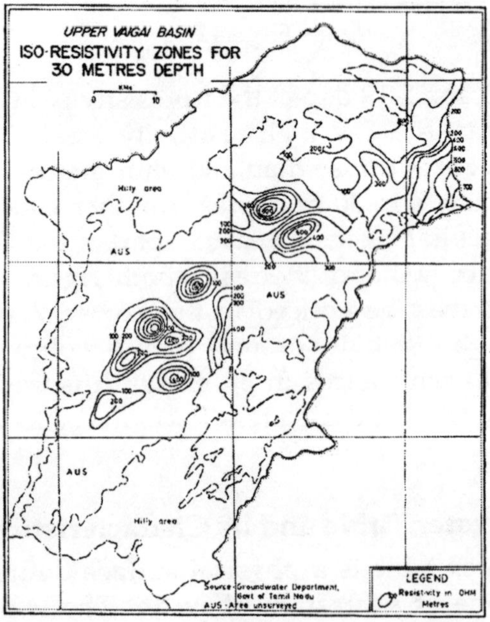

Figure 3.19

saturation and often fluctuates depending on the variations caused by addition or withdrawal of water. A rising groundwater table therefore registers an excess in the rate of recharge over and above the rate of discharge and vice-versa.

3.3.3.1 Location of Observation Wells

The characteristics of water table have been studied based on data from 35 observation wells (Fig. 3.20) selected on the basis of geologic conditions. The hilly areas have not been considered for the analysis due to the high altitude coupled with rugged topography. The monthly water table data have been collected for a period of thirty years from 1973 to 2002. An attempt is made to study the annual and seasonal groundwater table, annual and seasonal coefficient of variability of groundwater table, water table above mean sea level, and change in water table based on the available data.

Fig. 3.20

3.3.3.2 Mean Annual Groundwater Table

In the Upper Vaigai basin the depth of water table varies from place to place. Three areas of very deep water table are: the central part of the Upper Vaigai basin covering Lower Thalai Vaigai and Lower Suruliar sub-basin here the ground water table is located at great depth of more than 12 m (Fig. 3.21 and Table 3.12). In certain areas like Thadicheri and Thappukundu the water table much deeper (over 20 m). The second stretch of deep groundwater table is noticed near Sankarapuram in the Thevaram sub-basin where the depth is around 15 m from where the water table rises towards South in Uthamapalayam to a depth of 9 m; and lastly the third area of deep groundwater is located in the Koilpatti sub-basin where the water table is over 12 m in depth.

The depth of water table is in general at a minimum in the northeastern part and dips gradually towards the southwestern part of the basin. The water table is quite near the surface at a depth of less than 6 m from the surface in the Kottagudi sub-basin, foot hills of Lower Palanis, Andipatti hills and eastern part of the Viruveedu basin and the southeastern part of the Upper Suruliar sub-basin. In the Varushanad Valley the depth of the water table is over 9 m in the south 6 m to 9 m in the northern part of the valley. As already mentioned the depth of groundwater table is at its maximum of about 12 m in the southwestern part of the basin and gradually rises to about 6 m to 9 m towards the northeastern part of the basin (Fig. 3.21).

Figure 3.22 presents the mean annual groundwater table elevation above mean sea level. Water table has its highest elevation above mean sea level in the southwestern part of the basin where it is over 440 m. Gradually its elevation decreases towards the northeast of the basin. While crossing the Suruliar the water table makes a slight bend upstream (Fig. 3.22). In the Odaipatti micro basin and the Varushanad Valley the elevation of water table is 420 m MSL. The elevation of water table is almost uniform till it reaches the Lower Suruliar basin. In the Varahanadi basin the elevation is between 260 and 280 m MSL. The elevation in the northeastern part of the basin decreases towards the east. In the Lower Maruthanadi and the Viruveedu basins the elevation is less

Table 3.12
Upper Vaigai Basin: Seasonal and Annual Level of Water Table and Coefficient of Variability

Name of the Well	Winter		Summer		Southwest monsoon		Northeast monsoon		Annual	
	WT	CV	WT	CV	WT	CV	WT	CV	WT	CV
Kadamalaikundu	6.12	59.80	6.96	44.49	6.93	40.96	6.14	36.30	6.54	41.36
Gandamanur	5.12	60.65	7.69	35.57	11.51	30.25	8.53	30.64	8.21	32.89
Errasakkanayakkanur	7.60	57.20	9.13	47.64	10.25	47.64	10.02	31.83	9.25	40.65
Thadicherri	16.43	40.48	16.75	36.54	19.21	28.71	19.54	27.73	17.98	29.69
Venkatachalapuram	20.46	9.57	20.46	11.58	21.38	9.25	21.28	5.57	20.90	7.99
Kullappagoundanpatti	4.14	54.23	5.08	47.58	5.70	48.38	6.00	47.03	5.23	51.43
Melagudalur	13.85	4.63	14.31	23.81	14.31	34.28	14.25	5.63	14.18	4.60
Narayanathevapatti	10.42	17.83	12.12	18.83	13.76	14.00	12.36	15.08	12.17	15.13
Kambam	10.72	4.96	11.31	4.18	11.61	5.63	11.11	3.54	11.19	2.96
Seelayampatti	8.74	30.20	9.80	27.54	10.65	25.49	9.79	17.92	9.75	22.72
Kottur	9.32	36.81	1.24	28.83	12.54	19.75	12.27	29.29	11.34	26.07
Uthamapalayam	4.80	37.61	5.63	32.60	6.54	18.02	5.81	22.93	5.70	21.53
Mettupatti	12.14	19.97	12.91	17.79	12.98	19.25	12.97	18.75	12.75	20.05
Sankarapuram 1	14.08	19.71	14.04	21.71	14.60	19.60	14.82	18.73	14.38	19.33
Sankarapuram 2	15.87	27.85	16.89	26.10	16.91	21.51	16.81	18.60	16.62	20.79
Bodinayakkanur	5.13	42.09	5.91	23.95	7.13	8.02	6.61	14.11	6.20	15.41

Thiruchendur	3.21	55.84	4.22	55.75	4.66	41.91	4.00	76.72	4.02	34.14
Theni	4.24	21.71	4.90	18.45	5.68	16.73	4.91	17.24	4.93	15.09
Rathnam Colony	9.96	35.10	11.43	41.53	11.85	31.51	10.73	27.92	10.99	28.57
Arappadi Thevanpatti	8.55	40.97	10.81	32.41	13.08	23.24	11.46	28.83	10.98	25.00
Madurapurm	14.18	24.69	14.69	20.70	16.02	13.00	16.06	27.78	15.24	15.11
Vaigaipudur	8.55	35.78	10.81	30.15	13.08	13.23	11.46	16.31	10.98	10.07
Gandipuram	4.97	23.46	5.38	23.72	6.25	13.36	5.65	14.41	5.56	16.51
Gullapuram	5.83	2.93	7.14	38.03	9.42	17.43	7.69	34.34	7.52	21.76
Devadanapatti 1	5.00	27.40	5.08	31.73	6.37	29.33	6.14	22.22	5.65	20.04
Devadanapatti 2	3.59	34.99	3.92	41.76	5.42	40.90	4.52	36.07	4.36	35.75
Keelakoilpatti	5.56	45.76	7.49	39.80	9.62	34.13	6.08	34.53	7.19	28.99
Venkatasatrikottai	5.80	39.92	6.92	26.55	8.45	13.97	6.18	28.66	6.84	17.87
K. singarakottai	7.14	64.95	8.85	35.51	11.24	35.51	9.18	45.96	9.10	43.21
Viralipatti	6.55	51.29	8.12	44.26	10.65	29.01	7.58	32.22	8.23	30.94
Kottarpatti	4.78	63.09	6.60	42.13	8.51	19.88	5.63	33.08	6.38	30.48
Thimmarasanayakkanur	6.55	45.20	7.84	33.11	10.00	16.37	7.70	22.4	8.02	23.96
Onbathuveedu	4.02	23.62	4.42	27.11	4.86	22.74	3.86	10.85	4.29	22.12
Palacombai	4.28	49.88	5.01	45.27	6.66	33.54	5.54	36.56	5.37	37.09
Andipatti	6.51	35.67	8.27	38.15	10.31	27.48	8.36	26.73	8.36	29.02

WT – Water Table in metre
CV – Coefficient of Variability in percentage

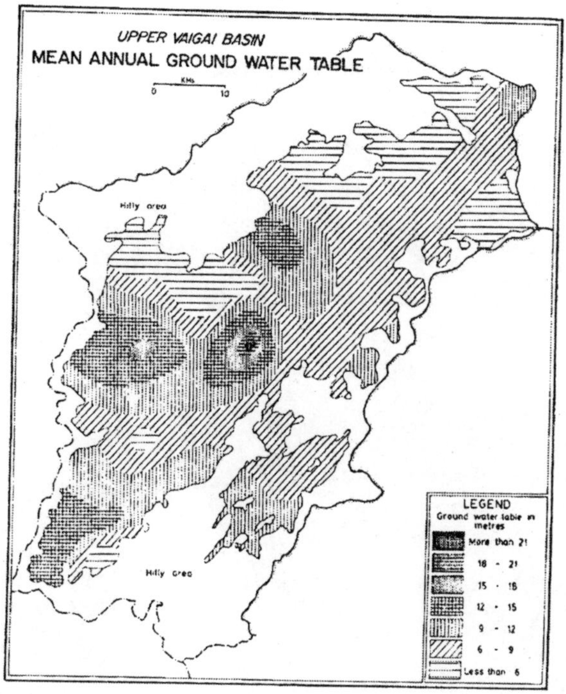

Fig. 3.21

than 240 m MSL. Thus the elevation of the water table above the mean sea level in general decreases towards north and northeast in concordance with the flow of the Vaigai River from southwest to northeast.

3.3.3.3 Mean Annual Groundwater Table: Trend Surfaces

Trend surface analysis defines the general pattern of mean annual groundwater table of the basin (Appendices 3.6 and 3.7). The liner trend surface equation fitted in is:

$$Z = 1688.442 - 13.027 X - 41.967 Y$$

The regional trend of the mean annual groundwater table is found to be a surface sloping down from the south-southwest to north-northeast (Fig. 3.23 and 3.24). In the south southwest the groundwater table is expected to be at depths of over 16 m each year while in the northeast it is expected to be about 6 m below the surface.

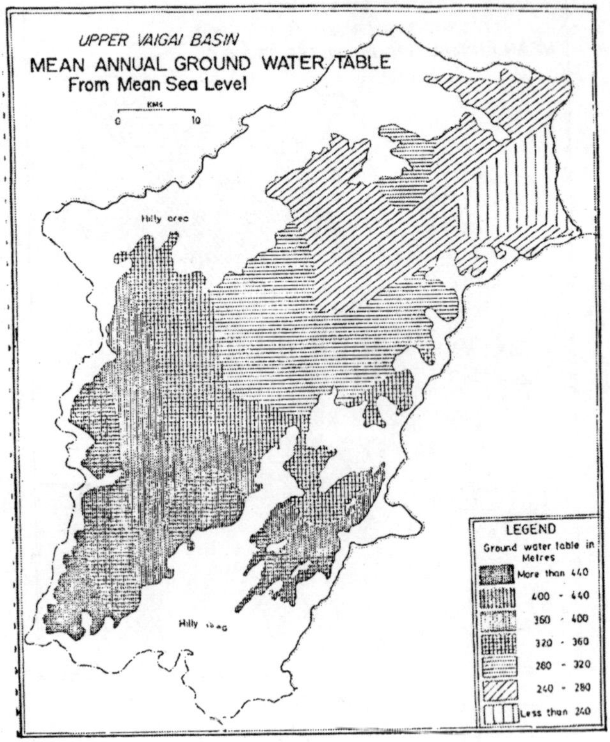

Fig. 3.22

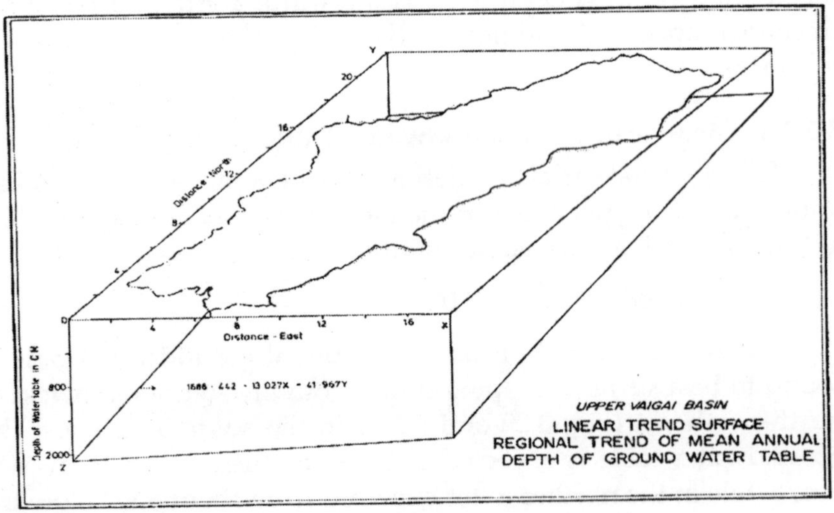

Fig. 3.23

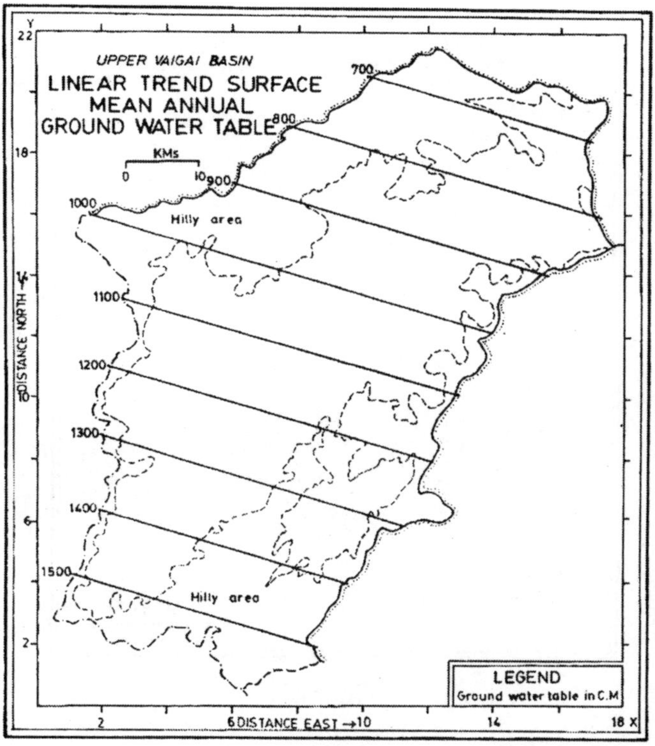

Fig. 3.24

A map residuals from the trend surface (Fig. 3.25) shows that the Thevarm, the Lower Suruliar, the Kottagudi, the Lower Varahanadi, the Upper Manjalar and the Upper Maruthanadi sub-basins stand out as areas with positive residuals whereas the Upper Suruliar, the Kottagudi, the Upper Varahanadi sub-basins and most of the eastern part of the study area are of negative residuals. The negative residuals have the water table close to the surface whereas the positive residuals coincide with the groundwater table at great depth. In the areas of positive residuals the prediction is correct to within a few centimeters of the depth of groundwater table; but in the Upper Suruliar, the Kottagudi and the Upper Varahanadi sub-basins the prediction has an error over 300 cm. The isopleths of residuals in Figure 3.25 indicate that these residuals follow a distinctive geographical pattern with two mounds in the central part of the basin.

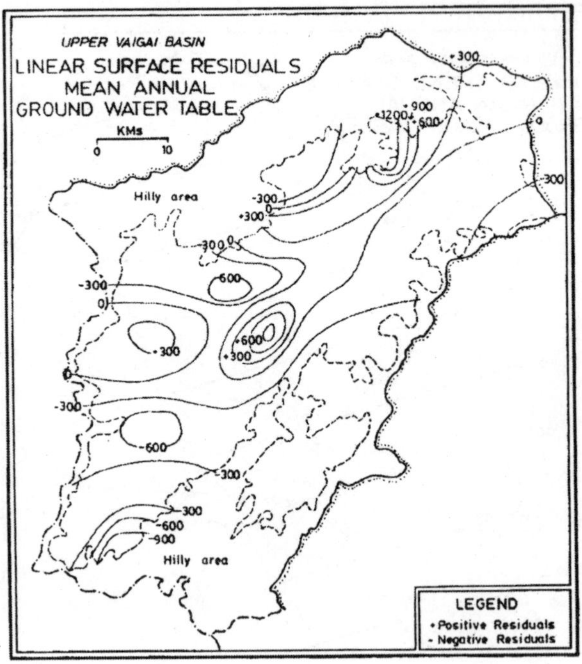

Fig. 3.25

3.3.3.4 Mean Seasonal Groundwater Table

The seasonal march of groundwater table is not very conspicuous (Fig. 3.26 and Table 3.12). Only during northeast monsoon period, there seems to be a slight rise in water table in the adjoining areas of the Lower Palanis and the southern part of the Upper Suruliar sub-basin, else where it remains almost the same with the no marked changes.

3.3.3.5 Variability of Groundwater Table

Level of variability indicates the level of stability. Lower the variability more is the stability and vice-versa. The annual and seasonal variability of groundwater table is presented in Table 3.12 and Figure 3.27. The coefficients of variability range from 8 to 51 per cent. It is high, over 50 per cent in the southeastern part of the Upper Suruliar sub-basin. From here, it gradually decreases towards west. The western part of the basin has the least coefficient of variability (less than 10 per cent) while in the Varushanad Valley

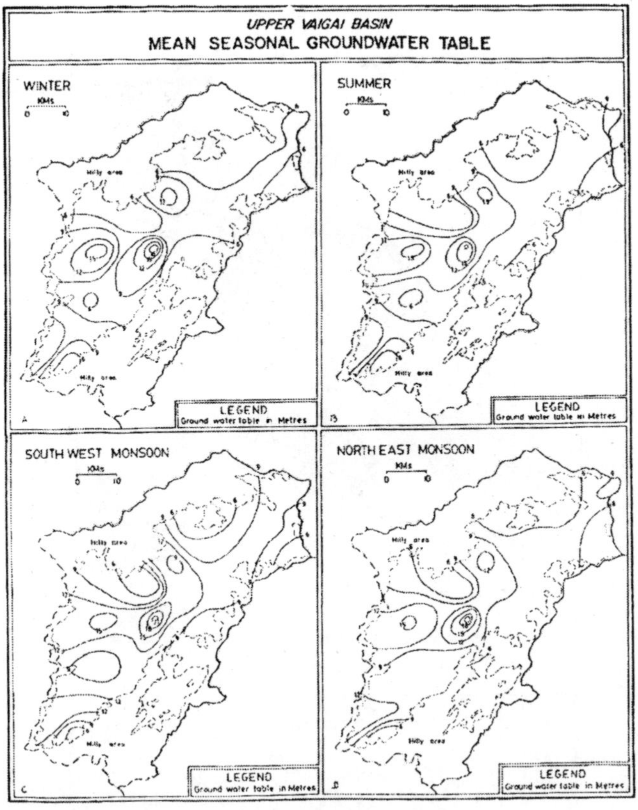

Fig. 3.26

and the Lower Thalai Vaigai sub-basin it is over 40 per cent. From the Lower Thalai Vaigai the variability decreases towards the north, northwest and west.

Low variability is recorded (20 per cent) in the Lower Varahanadi sub-basin (10.07 per cent). The Lower Suruliar and the Thalai Vaigai sub-basins also have low variability of water table.

The coefficients of variability of groundwater table for different seasons are depicted in Figure 3.28. Variability is very high during the northeast monsoon season. In the western part of the basin the variability is low in almost all the seasons and this shows the greater stability in the water table. The variability is high in the northeastern part and in the Varushanad Valley in almost all the seasons.

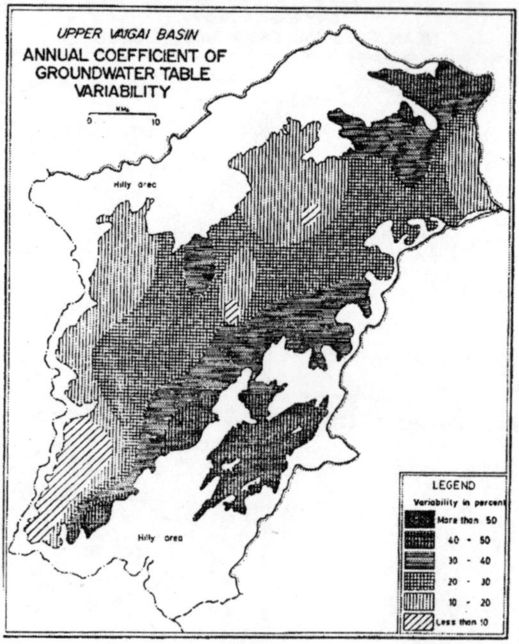

Fig. 3.27

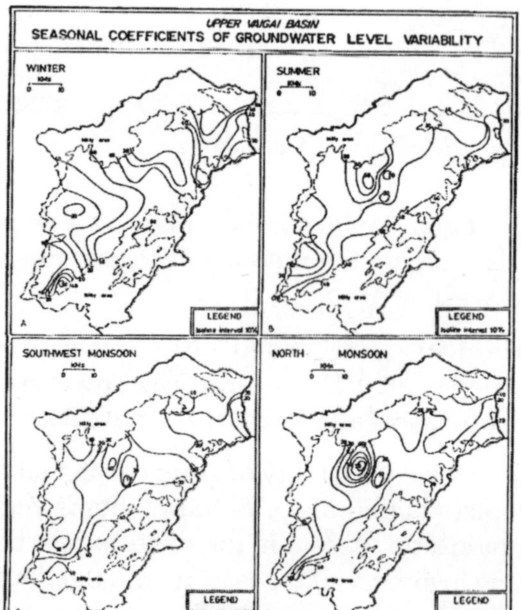

Fig. 3.28

3.3.3.6 Change in Groundwater Table

In this section an analysis has been made to study the change in water table after ten years. The annual water table of each well is calculated for then years and the overall change in water table is taken for the study.

In the western part of the Upper Suruliar basin the fall in the water table has been over – 1 m after a lapse of thirty years (Fig. 3.29). During the same period the water table has increased to + 1 m in the eastern side of the Upper Suruliar, the Lower Odaipatti, the Lower Thevaram and the Lower Suruliar sub-basins. In the Thevaram sub-basin covering the villages of the Kombai, T. Meenakshipuram and Thevarm and the western part of Pannaipuram the fall in water table is – 1 m while in Pottipuram village it has increased to + 1 m. In the Kottagudi and Lower Suruliar sub-basins the rise in water table is between 0 and + 1 m.

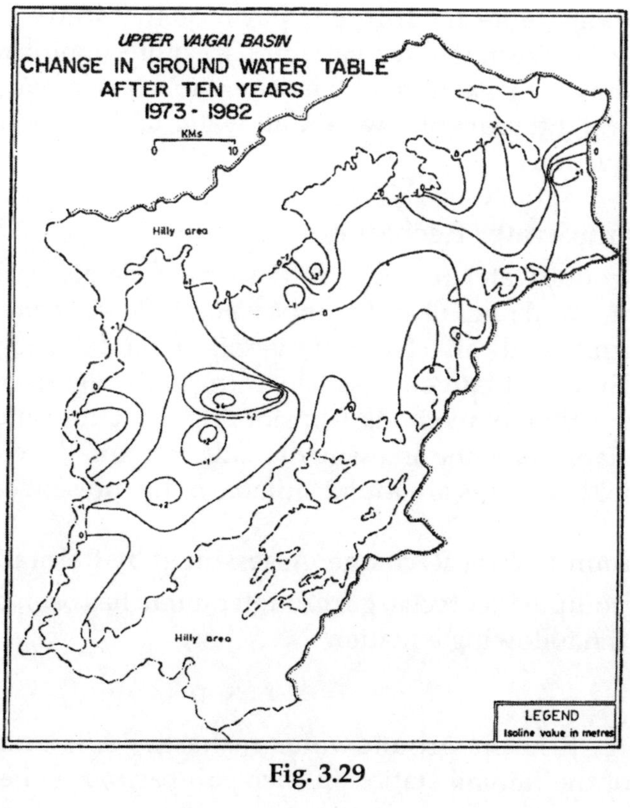

Fig. 3.29

A rise in water table up to 4 m has been observed in the Lower Suruliar basin especially in the areas of Uthamapalayam (+ 2 m), Markayankottai, Poolanadapuram (+ 2 m) and Kulaiyanur, Kottur and Thappukondu (+ 4 m). In the Varushanad Valley and Andipatti, Koilpatti and Lower Kottagudi sub-basins water table has risen between 0 and + 1 m. The Lower Manjalar and the Varahanadi sub-basins the water level has gone down by - 1 m.

Summing up, it could be said that the Lower Suruliar sub-basin stands out as a prominent area which has witnessed a positive rise in the water table during the last thirty years. The only other area which has witnessed a very slight rise is the northeastern extremity of the study area covering the Lower Maruthanadi sub-basin. Throughout the foot hill zone the water table has been lowered by about one metre during the same period.

Figure 3.30 reveals the relationship between rainfall and the level of water table in different parts of the study area. The relationship between the two variables are quite significant in that during years of normal rainfall the water table has remained stable while during the years of low rainfall water table fluctuations are quite conspicuous.

3.3.4 Groundwater Recharge

The groundwater recharge in the Upper Vaigai basin occurs mainly from local rainfall and groundwater inflow. The water table is being significantly affected by variations in rainfall and pumping of water. The quantitative assessment of groundwater recharge may be made by different methods. Considering the nature and amount of the available data, the assessment made by rainfall recharge and specific yield methods are found suitable in the present case.

3.3.4.1 Ground Water Recharge: Assessment by Rainfall

The groundwater recharge through rainfall has been calculated based on the following equation

$$R = A \times r \times p$$

Where, R is the groundwater recharge, A is area of each polygon of the rainfall station in each sub-basin, r is the recharge

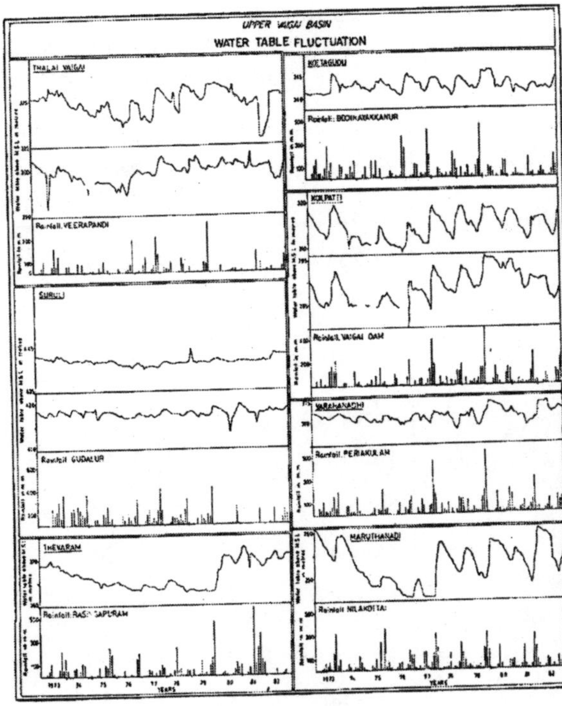

Fig. 3.30

percentage and p is the weighted mean rainfall of the each polygon. The recharge is then calculated for all the polygons in the sub-basin. The recharge percentage in the present calculation is adopted from the norms worked out by the Groundwater Division, Tamil Nadu (Public Works Department, 1976). This procedure for calculating the rainfall recharge is applied to each sub-basin of the study area.

The groundwater recharge through rainfall recharge in the Upper Vaigai basin is presented in Table 3.13. From this table it is observed that the total groundwater recharge is 291.855 million m^3. The Suruliar basin has more groundwater recharge of 63.974 million m^3 followed by the Thalai Vaigai 55.483 million m^3, the Thevarm 30.604 million m^3, the Maruthanadi 23.56 million m^3, the Varahanadi 23.531 million m^3, and the Andipatti 21.403 million m^3 sub-basins. In the Viralipatti sub-basin groundwater recharge is very low 9.439 million m^3.

Table 3.15
Upper Vaigai Basin: Ground Water Yield Through Specific Yield Method

Name of the sub-basin	Yield in million m^3	Percentage to the total yield
Thalai Vaigai	77.254	18.54
Suruliar	91.741	22.01
Thevaram	37.691	9.04
Kottagudi	17.505	4.20
Koilpatti	16.213	3.89
Varahanadi	58.569	14.05
Manjalar	25.004	6.00
Maruthanadi	25.109	6.02
Viralipatti	13.308	3.19
Viruveedu	15.144	3.63
Andipatti	39.195	9.41
Upper Vaigai Basin	416.733	100

Table 3.14 gives the return flow recharge and recharge in water spread of the basin. Based on the return flow, recharge is calculated. It is observed from the seepage studies conducted in different irrigated fields of the basin that seepage rate through return flow is taken as 90 cm and the seepage rate of 40 cm by water bodies (Public Works Department, 1981).

After calculating the return flow recharge and water bodies recharge, 15 per cent is deducted for the loss of evaporation and evapotranspiration. The remaining 85 per cent of the total recharge is taken as net recharge. Table 3.14 presents the total recharge through irrigation fields and water bodies of the basin. The total recharge through irrigation field is 228.456 million m^3. In the

Suruliar sub-basin the recharge is 90.036 million m^3. This is mainly because of the extensive irrigated areas by the canals and tanks. This is followed by the Kottagudi 34.342 million m^3, the Varahanadi 32.562 million m^3, the Manjalar 11.844 million m^3 and the Thalai Vaigai 11.889 million m^3 sub-basins. Very low return flow recharge is recorded in the Viruveedu, the Koilpatti and the Viralipatti sub-basins of 2.079, 2.079 and 0.072 million m^3 respectively since these basins are having negligible areas under irrigation and also due to the poor means of the irrigation sources.

With regard to recharge through water bodies the Koilpatti sub-basin leads 6.458 million m^3 as the Vaigai dam is located here. This basin is followed by the Suruliar 6.026 million m^3, the Varahanadi 4.928 million m^3, the Manjalar 2.669 million m^3, Maruthanadi 2.475 million m^3, the Thevaram 2.970 million m^3 and the Thalai Vaigai 1.882 million m^3 sub-basins. In other sub-basins a meager amount of water is added to the total recharge. The total recharge through all the water bodies is around 29.138 million m^3.

Table 3.15 highlights the quantity of groundwater recharge after loss through evapotranspiration. The rainfall recharge are clubbed together and from the total 15 per cent is reduced as allowance for the loss through evapotranspiration and the net recharge is given in the last column of the Table 3.18. The total recharge of the basin is 467.029 million m^3. In the Suruliar sub-basin the recharge is 136.030 million m^3 followed by the Thalai Vaigai 58.870 million m^3, the Thevaram 52.727 million m^3, the Varahanadi 51.867 million m^3 and the Kottagudi 43.857 million m^3 sub-basins. In the Manjalar, the Maruthanadi and the Koilpatti sub-basins the recharge is 26.180, 28.939 and 20.969 million m^3 respectively. The Andipatti, the Viruveedu and the Viralipatti sub-basins have recharge of 24.519, 14.828 and 8.245 million m^3 respectively.

3.3.4.2 Groundwater Recharge: Assessment by Specific yield Method

Based on the water level data, observed each month the mean groundwater fluctuation is calculated. The specific yield and water table fluctuations have been applied in solving the following equation to find out the groundwater recharge of the study area.

Table 3.14

Upper Vaigai Basin: Return Flow Recharge And Recharge In Water Spread Area

Name of the sub-basin	Irrigated area in ha	Return Flow recharge in million m^3	Water spread area in ha	Recharge in water spread area in million m^3
Thalai Vaigai	1321	11.889	419	1.887
Suruliar	10004	90.036	1339	6.026
Thevaram	3162	28.458	660	2.970
Kottagudi	3838	34.542	142	0.640
Koilpatti	231	2.079	1435	6.458
Varahanadi	3618	32.562	1095	4.81
Manjalar	1316	11.844	593	2.669
Maruthanadi	890	8.010	550	2.475
Viralipatti	8	0.072	42	0.189
Viruveedu	231	2.079	75	0.388
Andipatti	765	6.885	124	0.558
Upper Vaigai Basin	25384	228.456	6474	29.138

Table 3.13
Upper Vaigai Basin: Groundwater Yield

Name of the sub -basin	Recharge through rainfall	Return flow recharge	Water spread recharge	Total Recharge	Net Recharge after 15 per cent Evapotranspiration Losses
Thalai Vaigai	55.483	11.889	1.887	69.259	58.870
Suruliar	63.974	90.036	6.026	160.036	136.030
Thevaram	30.604	28.458	2.970	62.032	52.727
Kottagudi	16.415	34.542	0.640	51.597	43.857
Koilpatti	16.130	2.079	6.458	24.667	20.967
Varahanadi	23.531	32.562	4.928	61.021	51.867
Manjalar	16.287	11.844	2.669	30.800	26.180
Maruthanadi	23.561	8.010	2.475	34.046	28.939
Viralipatti	9.439	0.072	0.189	9.700	8.245
Viruveedu	15.028	2.079	0.338	17.445	14.828
Andipatti	21.403	6.885	0.558	28.846	24.519
Upper Vaigai Basin	291.855	228.456	29.138	549.449	467.029

Recharge = Area x Specific yield x Mean Water Table Rise

The areas of each polygon of the observation wells are arrived at first and the specific yield and rise in the water table in each polygon has been identified. Then the groundwater recharge for each sub-basin-wise groundwater recharge through specific yield method.

The total groundwater recharge of the basin is 416.733 million m^3. The Suruliar basin has huge quantities of groundwater amounting to 91.741 million m^3 followed by the Thalai Vaigai 77.253 million m^3, the Varahanadi 58.569 million m^3 and the Andipatti 39.195 million m^3 sub-basins. The Manjalar and the Maruthanadi sub-basins have 25.004 and 25.109 million m^3 respectively. In the basins of the Kottagudi, Koilpatti and the Viruveedu sub-basins a very low quantity of groundwater recharge 17.505 million m^3, 16.231 million m^3 and 15.744 million m^3 is noticed respectively. The lowest water recharge is found in the Viralipatti (3.340 million m^3), which has a very small basin with very low rainfall.

3.3.5 Groundwater Quality

The water quality determines the choice of crops and crop growth of any region. The periodic fluctuations of the water tables and the resultant changes in the chemical constituents of the groundwater are now realized. An attempt has been made in this section to analyze the spatial and temporal variations of groundwater quality in the Upper Vaigai basin. Analysis is made with the samples collected 35 observation wells.

The groundwater is classified into different types of cation and anion groups as in surface water quality analysis (vide section 3.2.5). The classification of this group is presented in Table 3.16 classification of the cation and anion group of waters are represented in Figure 3.31.

3.3.5.1 Classification of Groundwater: Cation Group

Figure 3.32A depicts the cation group of groundwater distribution. The western part of the basin covering the Thevaram and the Kottagudi sub-basins have the values of more than 50 meg

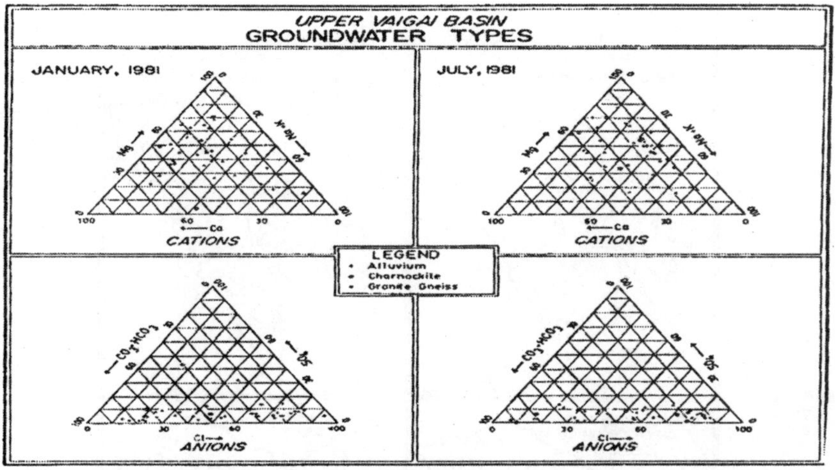

Fig. 3.31

per litre of cation group of waters. While in the Upper Suruliar sub-basin it is less than 30 meg per litre. In the Lower Thalai Vaigai the cation group of waters has more than 40 meg per litre. In the central part of the basic the cation group of ions ranges between 30 and 40 meg per litre. Among the cations group magnesium type of waters are found in the fourteen wells in the month of July and 10 in January. Magnesium ranks first in the cation group and calcium is very limited and is present only in January. Magnesium type of waters is distributed in almost the entire study area as they are found in all the geological formations. The total number of samples in his group gets reduced from 14 to 10 samples after the monsoons. The sodium waters are found mostly in granite gneisses and alluvium types. The calcium type of water is confined only in the month of January and is not seen in July. It is common to note that the intermediate type of waters is dominant among the water samples in the basin. The complexity of the rock types is the major factor responsible for this type of cation group of waters.

3.3.5.1 Classification of Groundwater: Anion Group

Figure 3.32B presents anion group of groundwater distribution. In the western part of the basin the anion group of groundwater is more than 40 mg per litre. The value exceeds 50 mg per litre in the

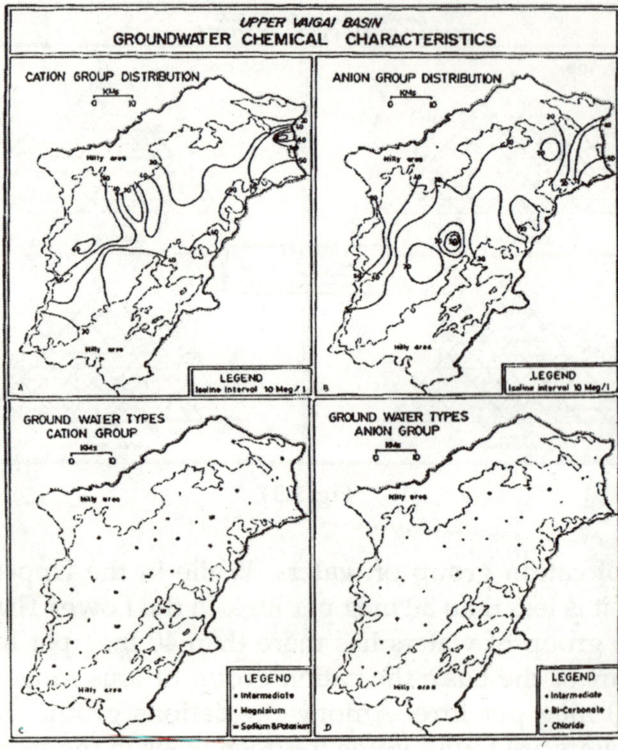

Fig. 3.32

Table 3.16

Upper Vaigai Basin: Groundwater Types

Groundwater Type	Number of Samples July '81	January '82
CATION GROUP		
Calcium	0	3
Magnisium	14	10
Sodium & Potassium	4	5
Intermediate	17	17
ANION GROUP		
Carbonate	16	14
Bicarbonate	0	0
Chloride	17	16
Intermediate	2	5

Lower Thalai Vaigai sub-basin. In the northeastern part of the basin the values exceed 40 mg per litre. In most of the southern and central part of the basin the anion group of waters has the values between 20 and 30 mg per litre.

Compared to cation group, anion group has a number of intermediate water samples. Chloride type of water occupies a majority of wells. Carbonate and bicarbonate type of water is mostly found in the granite gneiss type of formation. This type of water gets reduced after the monsoon rains. The chloride type of waters is found invariably in all the rock types. The distribution of intermediate group is very low when compared to other water types. In the premonsoon season 2 samples are seen whereas in January it has increased to 5 samples.

3.3.5.2 Salinity and Alkalinity

3.3.5.2.1 *Hydrogen Ion Concentration (pH)*

The pH values of groundwater samples of the basin range between 7.2 and 9 of this indicate that all the water types are alkaline (Table 3.17). The mean pH value of the groundwater in the basin is 7.8.

3.3.5.2.2 *Total Dissolved Solids (TDS)*

The distribution of the total dissolved solids is presented in Figure 3.33B. In the central part of the basin the TDS is more than 3000 mg per litre and is high in the southern part of the basin too where values of the TDS range between 500 and 1000 mg per litre.

3.3.5.2.3 *Electrical Conductivity (Ec)*

In the Upper Vaigai basin the average electrical conductivity in cation groups are as follows: Magnesium (1681 ppm) and intermediate (1636 ppm). Anion group of chloride has 2299 ppm, bicarbonate 1238 ppm and intermediate has 885 ppm. The average Ec of the groundwater types in the basin is 1711 ppm.

The distribution of electrical conductivity is presented in the Figure 3.33A. The values of Ec is high i.e. more than 4000 m mhos in the center covering the Lower Suruliar and the central Thevaram sub-basins. Another area with high value is located in the Lower

Table 3.17

Upper Vaigai Basin: Major Chemical Characteristics Of Surface Water

(values in equivalent parts per million)

Variables	Calcium type*	Sodium Type			Intermediate Type		
		Mean	Max.	Min.	Mean	Max.	Min.
Calcium	30.00	23.00	28.00	18.00	36.67	54.00	24.00
Magnesium	12.20	24.50	38.90	10.09	24.73	45.00	10.90
Sodium	9.20	75.90	94.30	57.50	50.09	92.00	9.20
Carbonate	-	51.00	90.00	12.00	35.33	180.00	12.00
Bicarbonate	134.00	229.00	244.00	214.00	281.89	372.00	116.00
Chloride	24.80	37.25	42.60	31.90	49.27	92.20	21.30
PH	7.90	8.25	8.30	8.20	7.96	8.20	7.70
TDS	173.00	342.50	465.00	220.00	341.67	550.00	115.00
EC	272.00	535.00	725.00	345.00	535.00	862.00	181.00
SAR	0.36	2.60	2.90	2.30	1.49	2.80	0.75
Total Hardness as CaCo3	125.00	160.00	205.00	115.00	193.33	320.00	105.00
Na in per cent	13.79	31.04	52.08	50.00	32.59	49.38	16.00
RSC	-	2.25	2.90	1.60	1.46	6.70	0.10

* Only one sample analyzed

Anion Group

Variables	Bicarbonate Type			Intermediate Type		
	Mean	Max.	Min.	Mean	Max.	Min.
Calcium	36.20	54.00	24.00	22.00	26.00	18.00
Magnesium	21.08	45.00	10.09	36.50	38.90	34.10
Sodium	42.55	87.40	9.20	93.15	94.30	92.00
Carbonate	27.00	180.00	12.00	75.00	90.00	60.00
Bicarbonate	232.90	372.00	116.00	241.00	244.00	238.00
Chloride	42.56	92.20	21.30	58.55	74.50	42.60
PH	7.96	8.20	7.70	8.20	8.30	8.10
TDS	295.20	550.00	115.00	490.50	516.00	465.00
EC	462.50	862.00	181.00	766.00	807.00	725.00
SAR	1.33	2.60	0.36	2.85	2.90	2.80
Total Hardness as CaCo3	177.50	320.00	105.00	205.00	205.00	205.00
Na in per cent	30.98	52.00	13.79	49.69	50.00	49.38
RSC	1.78	6.70	0.10	2.40	2.90	1.90

Andipatti sub-basin. The Upper Manjalar, the Maruthanadi, the Western Thevaram and the Viruveedu sub-basins have less than 1000 m mhos. In most of the central part of the study area the electrical conductivity ranges between 1000 and 2000 m mhos.

3.3.5.3 Sodium Adsorption Ratio

The sodium adsorption ratio is calculated based on the formula found in section 3.2.4.2. Water classification for SAR is as follows: with a value of more than 10 is excellent water. The values between 10 and 18 is good, between 18 and 26 is fair and more than 26 is poor water quality. Figure 3.33A depicts the distribution of sodium adsorption ratio of the basin. In the Varushanad Valley the values are more than 7. In the Upper Kottagudi basin has more than 5. The western part of the Upper Suruli basin has more than 4. The SAR decreases towards the south central part from the south of the basin. In most of the areas the SAR is between one and two.

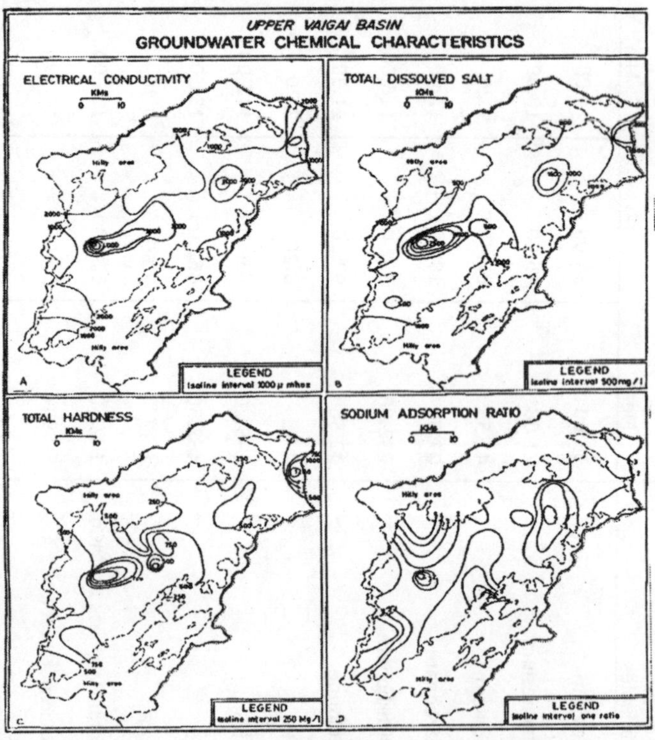

Fig. 3.33

3.3.5.4 Total Hardness of Water

Total Hardness is a measure of the calcium and magnesium content and is expressed as the equivalent of calcium carbonate. The distribution of total hardness is presented in Figure 3.33A. As seen in the previous distribution maps on water quality, a high concentration of more than 1000 mg per litre is noticed in the Lower Suruliar sub-basin. The total hardness varies from 500 to 750 mg per litre in the northeastern, central and southern part of the basin. The north central and northern part of the basin the total hardness varies from 250 mg to 500 mg per litre.

3.3.5.5 Classification of Groundwater

Groundwater may be classified on the basis of the dominant ions present. According to the classification of Richard (1954) Salinity (C) and Sodium (S) are divided into four groups low, medium, high and very high so that 16 water types can be distinguished varying from excellent (C_1S_1) to very poor (C_4S_4). In the eastern part of the Koilpatti sub-basin the classification is C_3S_1 and another patch in the Viruveedu sub-basin also belongs to this category. Most of the central and southern part of the basin the classification of groundwater is between C_3S_1 and C_4S_1. In the extreme southern part the classification is C_4S_1. In the northeastern part it is C_4S_1.

Change in quality of water results due to contamination and pollution by sewage and industrial water percolating into groundwater storage. Long term study of groundwater analysis is helpful in determining the changes in quality of water due to various reasons like contamination through domestic and municipal sewage and effluents form industries, degradation due to over development and use of irrigation water and percolation of surface drainage. No significant change in quality pattern has been noticed in the basin so far. Groundwater tapped along the river Vaigai locally shows slight chemical variation due to solvent action in weathered mantle to the induced infiltration from the river flow. A detailed survey conducted in each area revealed that the quality variation is a local phenomenon due to entry of domestic and municipal sewage. The suitability of groundwater in the basin is classified as good to permissible.

CHAPTER 4

WATER RESOURCES: DEVELOPMENT AND UTILIZATION

4.1 INTRODUCTION

An attempt to evaluate the surface and ground water resources of the Upper Vaigai basin has been elaborated in Chapter Three. This forms the basis for the study of development and utilization of water resources of the basin for comprehensive water resource planning. The availability, development and utilization of water resources are not the same in all the sub-basins of the study area mainly due to variation in their size and location and therefore a striking disparity in the economic development between various sub-basins. An attempt is made here to study the development of water resources and utilization of water for different purposes in the Upper Vaigai basin.

4.2 WATER RESOURCES DEVELOPMENT

The main purpose of water resources development is to achieve maximum economic and social returns. Development of irrigation and hydel power has been responsible in the economic and social uplift of the country due to its obvious benefits in agriculture and industries. Optimum utilization of the available water resources and augmentation through the development of water resources are important objectives of water resources development. Water resources development of any basin is based on the total availability of surface and ground water resources. For wise utilization of water resources, there is a need for the study of water resources development. The developmental programmes are for irrigation, community water supply and power generation. In this section an attempt is made to study the surface and ground water resource development of the Upper Vaigai basin.

4.2.1 Development of Surface Water Resources

Surface water resources remain the most important source of water for the basin requirements. Development projects of water resources, which are mostly multipurpose, aim at controlling river flow for the best advantage of the economy of a drainage basin (Bekinsale, 1969). Surface water is provided by stream flows, tanks or/and reservoirs. Irrigation from stream flow is mainly effected through canals drawn from dams constructed across the rivers. Tanks, another important source of irrigation, are mostly rain-fed and also depend on the replenishment of the surrounding drainage area. The Upper Vaigai basin has a long history in the development of surface water resources. As early as 1895, programmes were initiated to develop water resources.

4.2.1.1 Development Programmes and Benefits

The agricultural prosperity of the Upper Vaigai basin is largely due to the completion of the Periyar Irrigation Project. The waters of the Periyar, which flows down the western slopes of the Western Ghats across Kerala State, was diverted through a tunnel across the ghats for irrigating the parched lands of the then Madurai district. Although this idea was given as early as 1798 by Muthu Arula Pillai, the Prime Minister of the Raja of Ramnad, it was held up for want of funds. Shortly there after, in 1808 Captain Cald Well, the then district engineer of repute made a cursory examination, and felt that the scheme was not practical (Baliga, 1960). The matter, however, continued to be discussed until 1867, when it was brought forward by Major Ryves who proposed the construction of an Earthen dam 50 m high across the Periyar and send back the waters in a reverse direction through a cutting across the watershed. His idea was merely to divert the river and not to store the water, the estimated cost of which was more than 17.5 lakhs rupees. From 1868 to 1870 Colonel Penny Cuick and Smith investigated the scheme and a complete project estimated to cost 54 lakhs rupees was proposed.

Doubts arose as to the practicability of constructing so high an earthen embankment. However, Colonel Penny Cuick's proposal to lay a masonry dam was accepted in 1882 and he was directed to review the plan and estimate for the whole project. The scheme then drawn up included a great masonry dam across the Periyar, a huge lake and a tunnel through the watershed.

The scheme could be undertaken only after obtaining the consent of the then Travancore Government (now Government of Kerala), who had to agree to surrender not only the right of the waters of the Periyar which flow through their territory but also the right to some land on which the dam and the lake were to be constructed, but this consent was obtained only after prolonged negotiations form 1863 to 1886. Eventually by the agreement of 1886, the then Travancore Government granted on a 999 years lease to the former Government of Madras 32.38 km^2 land needed for the Periyar Project. The government also agreed to release 40 ha of land in the neighborhood for erecting temporary works and the right to carryout the Periyar works on these lands. The work began in 1887 and the project was opened in October, 1895 by Lord Wenlock, the then Governor of Madras State (Fig. 4.1) (Baliga, 1960). The Peranai regulator was constructed across the Vaigai River in 1898 in order to facilitate the distribution of water.

After utilizing the waters of the Periyar Irrigation Project for nearly sixty years it was felt that the Periyar waters were not fully utilized and consequently there was a considerable amount of runoff from the Periyar to the Arabian Sea. It was also felt that the fall of waters 366 m below the exit of the tunnel was not properly utilized as the bulk of the water was being wasted. The supply of water in the lower reaches of Vaigai is both precarious and scanty due to uncertain rainfall, and consequently the agricultural operations were seldom rewarded by a good crop. To overcome this uncertainty it was contemplated to use the surplus waters of the Periyar. As it was not a practical proposition to increases the storage capacity of the Periyar Lake the Vaigai Project was conceived in 1958 to conserve the surplus flows of the Periyar for the development of irrigation, besides generation of hydel power. Thus the Periyar Power House was started at the foot hills of the Western Ghats near Lower Camp in 1958 (Fig. 4.2).

UPPER PERIYAR & VAIGAI BASINS
LOCATION OF DAMS & ANICUTS AND DISTRIBUTION OF CANALS
KMs
0 10
TUNNEL
LEGEND
Dam
Anicut
Canal
Stream

Fig. 4.1

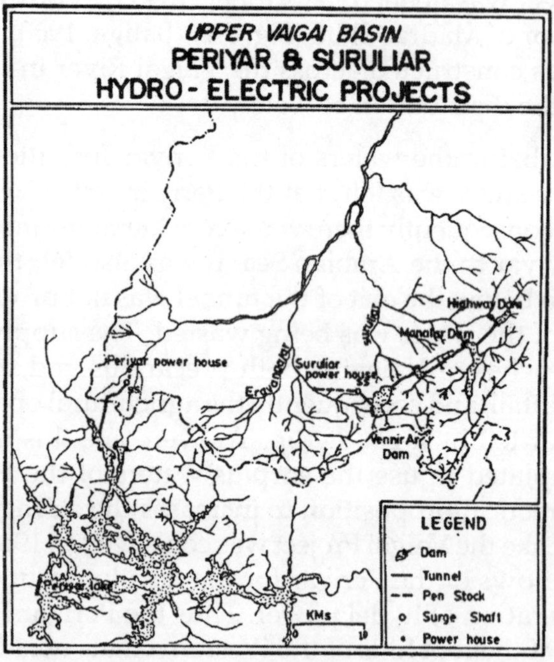
UPPER VAIGAI BASIN
PERIYAR & SURULIAR
HYDRO - ELECTRIC PROJECTS
Highway Dam
Manalar Dam
Periyar power house
Suruliar power house
Vennir Ar Dam
Periyar lake
KMs
LEGEND
Dam
Tunnel
Pen Stock
Surge Shaft
Power house

Fig. 4.2

After Independence, various medium and minor irrigation schemes were developed. The Manjalar and Maruthanadi dams were conceived during the last decade. Due to various efforts taken by the state government and local agencies, the development of tank irrigation was accelerated in different sub-basins of the study area. Various major and minor drinking water supply schemes for rural and urban areas have also been carried out in the basin.

4.2.1.2 Spatial Pattern of Surface Water Development

Surface water development has been mainly in the form of irrigation schemes. These schemes are broadly classified into major, medium and minor schemes on the basis of the area irrigated. The major schemes irrigate over 10,000 ha; medium schemes, 2000 to 10,000 ha; and minor schemes, less than 2000 ha. (Planning Commission, 1978).

4.2.1.2.1 Major Irrigation Projects

The agricultural prosperity of the Upper Vaigai basin has been considerably improved after the completion of the Periyar Irrigation Project. The Periyar dam was constructed at a height of about 47 m across the River Periyar in order to raise the water level of the Periyar Lake. The waters from this lake is taken to the eastern slopes of the Western Ghats by the Mulaipuzha Valley situated on the north of Periyar Lake through an open cut, to a length of 1860 m and then taken through a tunnel 1740 m long (Fig. 4.1). From the mouth of the watershed of the Periyar Lake through the tunnel the water passes by the Vairaravanar, a tributary of the Suruliar into the main Suruliar. To enhance the irrigation water in the Suruliar basin, 13 small anicuts were constructed at various points along the course of the river (Fig. 4.3).

An extent of about 5840 ha is being irrigated in the Suruliar basin. The entire area is under double crop. Normally every year the canals are left open in the first week of July and closed in the first week of March. The network formed by the canals is quite dense in the Suruliar basin (Fig. 4.3). Some of the important canals of this basin are the Palayamparavu, the Chinnamanur, the Seelayampatti, the Kottur, the Upparapatti and the Veerapandi (Vide Appendix 4.1). The ayacut under these canals varies from 154 to 724 ha.

Chinnamanur canal irrigates the largest segment of this basin and Peyathevan canal, the smallest. P.T. Rajan canal starts from the Uthamuthu Anicut and connects the canals of the Paalyampararu, the Karunkattankulam, the Valasamudram tank, the Sokki Odai and three newly formed tanks.

Fig. 4.3

The Vaigai dam has been put-up across the river Vaigai in 1958 near Andipatti (Fig. 4.1) to facilitate the drawl of water from the Periyar Dam all through the year and to store the flood waters of Vaigai. This project resulted with the following additional benefits: an additional extent of 9227 ha has been added to the ayacut; an assured supply to the existing tanks fed by the Vaigai, irrigating 54634 ha; A quantity of 17 million m^3 of water had been set apart for release for the Madurai Corporation water supply. The Peranai regulator was constructed across the Vaigai River at a distance of 39 km from the Vaigai Dam in Mathagupatti village of Nilakottai taluk (Fig. 4.1) in 1898. (Public Works Department, 1898).

It has two canals, the Periyar Main Canal and the Thirumangalam Main Canal. The Periyar Main Canal is used for irrigating about 12990 ha of double crop whereas the Thirumangalam Main canal irrigates 5020 ha of single crop. The ayacut of this regulator is situated outside the Upper Vaigai basin in Nilakottai and Melur taluks of Anna and Madurai districts respectively. The main work of Peranai is to regulate the water to these canals.

4.2.1.2.2 *Minor Irrigation Projects*

Next in importance are the minor irrigation projects in the upper Vaigai basin. There are two important projects Manjalar and Maruthanadi Irrigation Projects. The other minor irrigation schemes in each sub-basin are being dealt separately in this section.

4.2.1.2.2.1 *Manjalar Irrigation Project*

The Manjalar Irrigation Project is located in the Manjalar sub-basin. The dam was constructed across the Manjalar near Devadanapatti in Periyakulam Taluk (Fig. 4.1) in the year 1967. The catchment area of the dam is 120 km^2. The total capacity of the dam is 13.792 million m^3. A canal 7.25 km long along the right side of the river course irrigates an area of about 810 ha. All along the course of the river nine anicuts are constructed (Fig. 4.1).

4.2.1.2.2.2 *Maruthanadi Irrigation Project*

The Maruthanadi Irrigation Project is located in the north eastern part of the basin (Fig. 4.1). A reservoir with a capacity of 5.338 million m^3 is constructed across the Maruthanadi River near Ayyampalayam village in Dindigul taluk. The catchments area of the river at the dam site is 33.15 km^2. The left main canal runs for a distance of 10 km and provides water for 739 ha of cropped land and 288 ha of tree crops in the Lower Maruthanadi basin. The right main canal, which has a length of 10 km, has an ayacut of 495 ha of land and 158 ha of tree crops. The area benefited by this project is about 1680 ha.

4.2.1.2.2.3 *Thalai Vaigai basin*

The main scheme in the Thalai Vaigai sub-basin is P.T. Rajan Channel Scheme. A major portion of the registered ayacut under the tanks in Venkatachalapuram was practically dry. Hence it was proposed to augment the supply of water to these tanks by the Periyar waters from the Suruliar basin.

The P.T. Rajan channel draws water from the Suruliar basin above the Uthamuthu anicut through the existing Palaymparau canal into Karunkattankulam (Fig. 4.4). From this tank a new canal leads to Valasamudram tank, irrigating the direct ayacut in transit. The balance quantity of water is drawn from Valasamudram tank through Sukkamodai to the various existing tanks and also to the three newly constructed tanks. The areas benefited are in Poomalaikundu, Thadicheri and Thappukundu villages in the Thalai Vaigai basin. The total irrigated area is 54.3 ha. (Public Works Department, 1982). Other minor irrigation schemes in this basin are Narasingapuram scheme and Vallalnadi Scheme, which irrigate an area of 251 ha and 317 ha respectively.

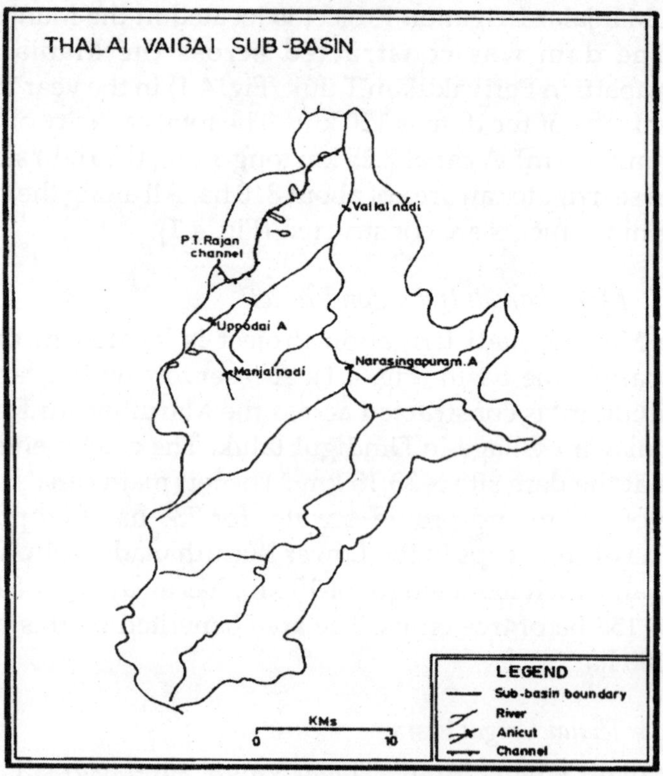

Fig. 4.4

4.2.1.2.2.4Thevaram Sub-basin

With regard to the Thevaram sub-basin, about 850 ha of land are being irrigated by the minor irrigation schemes. There are seven anicuts and five canals in this minor project area. Ammakulam canal which accounts for 91 per cent of the ayacut of the basin is the most important in this sub-basin.

4.2.1.2.2.5 Kottagudi sub-basin

Though there is no major dam in this basin, various anicuts have been constructed all along the course of the Kottagudi River in order to enhance and regulate the Water Supply (Fig. 4.5).

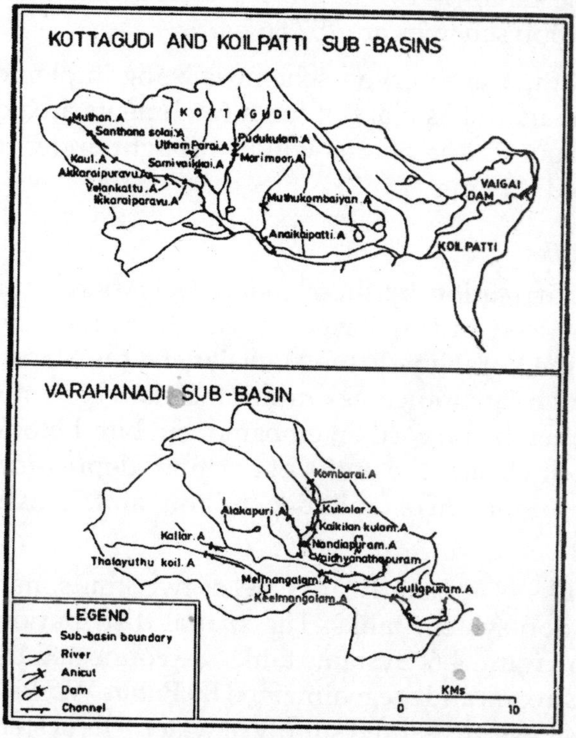

Fig. 4.5

4.2.1.2.2.6 Varahanadi Sub-basin

The Varahanadi has its source from the Berijam Lake (2164 m above MSL) in the Lower Palanis 19 km south of Kodaikanal. Due to the pressing demand for assured supply of water for agriculture the Varahanadi project came into being in the year 1869 (Public

Works Department, 1898). About 1800 ha are irrigated in this basin. The Berijam Lake supplies the domestic water needs of Periyakulam town too.

Apart from this project, there are a few other anicuts all along the course of the Varahanadi. The Thalayuthu Koil Dam, the Melmangalam anicut and the Keelmangalam anicut are constructed across the Varahanadi. The Kollar, the Kombarad, the Alagapuri, the Kaikilakulam and the Kukkalar anicuts are constructed across the tributaries of the Varahanadi (Fig.4.5). The Thalaiyuthu Koil Canal, Melmangalam Canal, Kutmangalam Canal and Gullapuram Canal are the canals in this basin. The total area irrigated by these minor irrigation schemes are 757.86 ha.

The Sothuparai reservoir scheme is being implemented in the Upper Varahanadi basin and it has a catchments area of 48.64 km^2 (Fig.4.5). About 853 ha of area will be brought under irrigation by this proposed scheme.

4.2.1.2.3 *Tanks*

Besides irrigation by these canals, irrigation by tanks is also widely practiced in this basin, especially in the Thevaram, the Kottagudi, the Varahanadi, the Manjalar and the Maruthanadi sub-basins. Tank irrigation comes under the class of minor irrigation. Tank is formed by an earthen embankment bund blocking the line of a stream or closing the outlet of a natural depression in which it collects. Development of tanks depends on rainfall distribution and topography.

The tanks of the study area are of two kinds, namely system tanks and non-system tanks. The spatial distribution of tanks is depicted in Figure 4.6. Systems tanks are connected by canal of an anicut or a stream, and are maintained by Public Works Department. These tanks have an assured supply of water. The system tanks have a perennial problem of silting and it is necessary to have periodic desilting so as to maintain the volume of flow.

In the Suruliar basin, tanks are interconnected with canals. Of the total area, 1049 per cent is irrigated by the tanks in the Suruliar basin. The non-system tanks have separate catchments and they are full only during the rains and dries up later as there

Fig. 4.6

are no feeders such as canals or streams. These tanks have their own ayacut, which is limited in extent. The small non-system, tanks, which command less than 40 ha of ayacut are maintained by the respective Panchayat Union. The area irrigated by non-system tank is insignificant.

4.2.2 Development of Ground Water Resources

The development of surface water resources has not been found to cope with the increased demand. With the result, emphasis has now shifted to the exploitation of ground water resources. Consequently well irrigation is receiving greater attention.

Irrigation schemes of groundwater have been receiving greater attention with the inception of the Five Year Plans which emphasized the need of small farmers. The tempo of progress of the groundwater scheme was specially accelerated during the period 1966-69, thanks to green revolution and rural electrification programmes. There has been a tremendous increase in the use of groundwater, because of the relative reliability and little possibility for further exploitation of surface water.

Groundwater is tapped through wells which may be open or tube. The distribution of wells in any region is controlled by geological formations and topography and also the demand. In the study area, there are about 29792 wells and wells are found concentrated along the river courses. The concentration is conspicuous in the areas of Vattlagundu, Bodinayakkanur, Palanichettipatti and Periyakulam, where the water table is quite high.

Well irrigation is conspicuous in the Upper Vaigai basin, which accounts for 70 per cent of the area irrigated in the basin. Regional variations are also quite prominent although almost in all sub-basins expect in the Varahanadi and the Manjalar. Well irrigation accounts for over 50 per cent of the total irrigated area whereas in the Varahanadi and Manjalar it is 45 and 49 per cent respectively. In the Thalai Vaigai, Thevaram, Koilpatti, Viralipatti, and Andipatti well accounts for 90 to 98 per cent of the total irrigated area. This clearly explains the exploitation of groundwater of agriculture.

4.3 WATER RESOURCES UTILIZATION

The availability of water is limited and demand for it has been increasing rapidly and as such there requires an efficient utilization of water. This section outlines the pattern of present utilization of water in the basin.

4.3.1 Utilization of Water Resource for Agriculture

The dependence of the agrarian economy on the availability of water needs no emphasis. Utilization differs according to the crops grown. Before the advent of the major and minor irrigation project the Upper Vaigai basin cultivated dry crops although at present both wet and dry crops are cultivated. In the sub-basins of the Suruliar, the Manjalar and the Maruthanadi which have major irrigation projects wet crops such as paddy and bananas dominate whereas in the other sub-basin namely Thalai Vaigai, Andipatti, Viralipatti and Viruveedu dry crops such as Cholam, Kambu, Ragi, Samai and groundnut dominate. In the Suruliar sub-basin double crops are raised with the help of irrigation whereas in some other sub-basins such as the Kottagudi, Varahanadi and Manjalar only a single crop is raised.

4.3.1.1 Utilization of Surface Water Resources

Surface water utilized through canals and tanks are considered for the analysis of surface water utilization. 41.7 per cent of the total geographical area of the Upper Vaigai basin is the cropped area. Of this the net sown area is 36.2 per cent and area cropped more than once is 5.5 per cent. That is 15.53 per cent of the net sown area is cropped more than once. The irrigated area accounts for 32.98 per cent of the total cropped (Table 4.1) and the canal and tank irrigation account for 17.41 and 12.3 per cent of the total irrigated area respectively. Thus only 29.7 per cent of the total irrigated area is irrigated by surface water.

Among the areas irrigation by surface water, canal irrigation is the mot prominent accounting for 58.6 per cent followed by tank irrigation accounting for the remaining 41.4 per cent (Table 4.2). The maximum area irrigated by canals is in the Suruliar basin accounting for 77.80 per cent of the total area irrigated by surface water in this basin and nearly 50 per cent of the area irrigated by canals in the entire Upper Vaigai basin. This is followed by Manjalar where 76.65 per cent of the area irrigated by surface water by canals, the Kottagudi 51.69 per cent, the Varahanadi 46.00 per cent and the Maruthanadi 40.68 per cent. Although canals form the important source with reference to the area irrigated by surface water the total area irrigated by canals is not significant when compared to the entire Upper Vaigai basin. Varahanadi sub-basin has only 15.4 per cent of the canal irrigated area of the Upper Vaigai basin followed by Kottagudi 13.92 per cent, Manjalar 12.34 per cent and Maruthanadi 6.95 per cent.

In the Thalai Vaigai sub-basin canal irrigation contributes only 23.40 per cent of the total area irrigated by surface water and 1.6 per cent of the total area irrigated by canals in the Upper Vaigai basin.

The Koilpatti, Thevaram, Viralipatti and Andipatti sub-basins are conspicuous by the absence of canal irrigation while tank irrigation accounts for the entire surface irrigation. In the Viruveedu sub-basin 97.83 per cent of the total area irrigated by surface water is by tanks. This is followed by Thalai Vaigai 76.60 per cent, the Maruthanadi 59.32 per cent, the Varahanadi 54 per cent and the Kottagudi 48.31 per cent. Due to the dominance of canals, tank

Table 4.1
Upper Vaigai Basin
Total Cultivated Area and Total Irrigated Area

Name of the Sub-basin	Total area cultivated	Area irrigated	Percentage to the total cultivated area
Thalai Vaigai	28772	8601	29.89
Suruliar	39484	12990	32.90
Thevaram	17208	6225	36.18
Kottagudi	16658	5601	33.62
Koilpatti	4711	1464	31.07
Varahanadi	21685	5831	26.89
Manjalar	10605	3055	28.88
Maruthanadi	8825	4387	49.71
Viralipatti	1380	587	42.54
Viruveedu	6269	2019	32.21
Andipatti	10248	3943	38.48
Upper Vaigai Basin	165845	54703	32.98

irrigates less than a fourth of the area irrigated by surface water in the Suruliar and the Manjalar sub-basins.

Of the total area irrigated, the area irrigated by surface water is more in the Suruliar basin which accounts for over a third (37.55 per cent) followed by the Varahanadi sub-basin, where 19.5 per cent is by canals and tanks. In the Maruthanadi basin 1637.40 ha of land is irrigated by surface water, which accounts for 10.01 per cent of the total irrigated areas. In the remaining sub-basins the area irrigated by surface water is negligible.

Based on the data collected for 1999-2002 the area irrigated by surface water has been calculated (Table 4.2). On an average 0.90 ha. m of surface water is used for irrigating one hectare of land (Nag and Kathpalia, 1975). The surface water utilization has been computed using this quantity. The mean annual surface water utilization is 147.158 million m^3 that is 46.87 per cent of the total yield of surface water (Table 4.2). Of this the Suruliar basin utilizes

Table 4.2
Upper Vaigai Basin: Surface Water: Irrigated Area, Yield and Utilization

Name of the sub-basin	Irrigated area in ha	Canals in per cent	Tanks in per cent	Yield in million m^3	Utilization in percent	Balance in per cent
Thalai Vaigai	667	23.39	76.61	57.326	10.48	89.52
Suruliar	6139	77.80	22.20	52.484	105.25	-
Thevaram	323	-	-	10.203	28.50	72.50
Kottagudi	2581	51.69	48.31	49.338	47.08	52.92
Koilpatti	93	-	100.00	6.406	13.11	86.79
Varahanadi	3191	46.00	54.00	68.692	41.80	58.20
Manjalar	1542	76.65	23.35	19.969	69.50	30.50
Maruthanadi	1637	40.68	59.32	22.819	54.58	33.42
Viralipatti	22	-	100.00	3.764	5.37	94.63
Viruveedu	46	2.17	97.83	11.730	3.56	96.44
Andipatti	108	-	100.00	11.211	8.64	91.36
Upper Vaigai Basin	1635	58.61	41.39	313.952	46.87	53.13

one third of the total surface water of the basin. This is followed by the Varahanadi 19.51 per cent, the Kottagudi 15.79 per cent, the Maruthanadi 10.01 per cent and the Manjalar 9.43 per cent sub-basins.

The Thalai Vaigai basin utilizes 4.08 per cent of the total surface water utilized in the study area. A very small quantity of surface water of less than one million m^3 is used in Andipatti (0.66 per cent), Koilpatti (0.57 per cent), Viruveedu (0.28 per cent) and Viralipatti (0.14 per cent) sub-basins.

4.3.1.2 Utilization of Groundwater Resources

Table 4.3 presents the groundwater availability and utilization in different sub-basins of the study area. Area irrigated by groundwater is presented in this Table. Of the total irrigated area in the Upper Vaigai basin well irrigation contributes 70.30 per cent. One fifth of the area irrigated by wells is in the Thalai Vaigai basin where the rainfall is only between 800-900 mm per annum and the variability as high as 20 to 30 per cent, which makes rainfall unreliable. In addition the undulating topography has given rise to the predominance of well irrigation in these regions. Suruliar basin has only 17.73 per cent of the total area irrigated by wells in the Upper Vaigai basin. Although canal irrigation predominates here, groundwater is used a supplement for canal irrigation. The Thevaram basin account for 16.05 per cent of the total irrigated area under wells.

A similar feature is seen in the Thalai Vaigai basin and in the Andipatti sub-basins. Since there is no permanent supply of surface water sources well irrigation predominates. The Kottagudi, the Maruthanadi and the Varahanadi sub-basins have 7.81, 7.11 and 6.83 per cent respectively to the total area under well irrigation in the Upper Vaigai basin. Mention should be made of the insignificance of wells as sources of irrigation in the Viruveedu, the Manjalar, the Koilpatti and the Viralipatti sub-basins.

Based on the data collected from the village records the area irrigated by groundwater is calculated using 1979-82 records (Table 4.3). At present, on an average 0.65 ha. m of groundwater is used for irrigating one hectare of land (Nag and Kathpalia, 1975). The groundwater utilization of the basin is computed using this value.

Table 4.3

Upper Vaigai Basin: Groundwater : Irrigated Area Yield And Utilization

Name of the sub-basin	Well irrigated area in ha	Yield in million m^3	Utilization in per cent	Balance in per cent
Thalai Vaigai	7933	58.870	87.59	12.41
Suruliar	6851	136.030	32.73	67.27
Thevaram	6202	52.727	76.46	23.54
Kottagudi	3020	43.857	44.76	55.24
Koilpatti	1371	20.967	42.50	57.50
Varahanadi	2640	51.867	31.81	68.19
Manjalar	1513	26.810	37.57	62.43
Maruthanadi	2749	28.939	61.75	38.25
Viralipatti	565	8.245	44.54	55.46
Viruveedu	1973	14.828	86.49	13.51
Andipatti	3835	24.519	101.67	-1.67
Upper Vaigai Basin	38652	467.029	53.66	46.34

The mean groundwater utilization per annum is 250.59 million m^3 in the Upper Vaigai basin. The utilization the Thalai Vaigai sub-basin is one fifth of the total groundwater utilization in the basin. The Suruliar sub-basin accounts for 17.77 per cent of the groundwater utilized in the Upper Vaigai basin followed by the Thevaram 16.09 per cent, the Andipatti 9.95 per cent, the Kottagudi, 7.83 per cent, the Varahanadi, 7.13 per cent and the Maruthanadi 6.59 per cent sub-basins. The utilization of groundwater is the least in the Viralipatti sub-basin which accounts for 1.47 per cent of the groundwater utilized in the study area (Table 4.3).

4.3.2 Spatial Pattern of Domestic Water Supply

4.3.2.1 Rural Water Supply

The drinking water supply sources of in the rural areas of the basin are tanks, canals and wells which are also used for irrigation. The tanks and canals are seasonal. Wells have been a comparatively better source, but, due to the low rainfall, they have not be dug very deep. Drinking water supply sources of the rural areas of the basin are classified according to the availability of water. They are hamlets with no source of water within the hamlet; hamlets where the source yields only non-potable water; hamlets where water is potable but source is non-perennial; hamlets where water is potable and perennial but the source is either private or unprotected*; hamlets where there is no good source within the hamlet but an alternative source is available within one kilometer, and hamlets where there is good water sources available.

Table 4.4 presents type-wise distribution of rural drinking water availability of the Upper Vaigai basin. The basic unit for the study is hamlet or a group of more than 100 houses. The spatial distribution of rural drinking water availability is presented in Fig. 4.4. In all about 638 hamlets are distributed among 129 villages of the Upper Vaigai basin. The maximum number of hamlets is found in the Thalai Vaigai which accounts for 23 per cent of the total. Of these 39 per cent (57 hamlets) have good sources of drinking water within the hamlets and thirty per cent have potable water which is either private or unprotected. These two categories of hamlets

* For categories 1 to 4 there is no alternative good source available within a distance of one kilometer.

are found along the low land stretch of the river. Nearly 20 per cent have potable water from non-perennial source mostly in the rain-shadow regions of the basin in the Varushanad foot hills. Eleven per cent of the hamlets have no potable water and no source of drinking water and these hamlets are in the Varushanad and Andipatti hills.

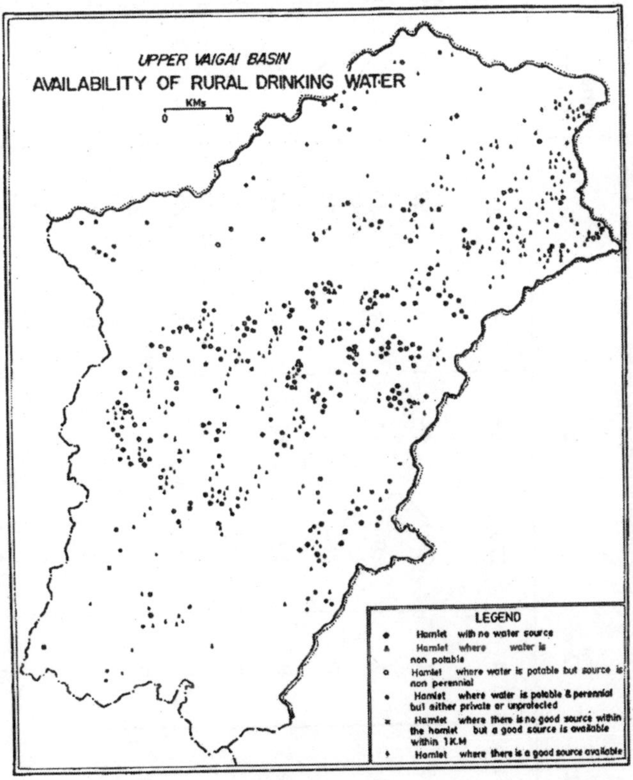

Fig. 4.7

The next important sub-basin from the point of view of distribution of hamlets is the Suruliar accounting for 13.2 per cent of the total hamlets of the Upper Vaigai basin. Among these hamlets 56 per cent have good sources of drinking water mainly along the banks of the river in the low lands. Nearly 23 per cent of the hamlets along the foot hill zone have potable water from non-perennial sources. This is mainly due to the rain-shadow effect in the southwestern part of the Suruliar basin. In between the two areas

Table 4.4
Upper Vaigai Basin: Type-wise Distribution Of Rural Drinking Water Availability

Name of the sub-basin	Type 1	2	3	4	5	6	Total Hamlets	Total Revenue Villages
Upper Vaigai Basin	21	4	176	93	5	339	638	129
Thalai Vaigai	16	1	29	44	0	57	147	19
Suruliar	1	1	19	13	3	47	84	24
Thevaram	1	0	26	6	0	34	67	11
Kottagudi	1	0	10	13	2	12	38	10
Koilpatti	1	2	18	4	0	14	39	6
Varahanadi	0	0	10	7	0	28	45	14
Manjalar	0	0	3	1	0	26	30	7
Maruthanadi	1	0	7	4	0	51	63	14
Viralipatti	0	0	6	1	0	12	19	2
Viruveedu	0	0	4	0	0	32	36	10
Andipatti	0	0	44	0	0	26	70	12

already discussed are distributed nearly 16 per cent of the hamlets of the basin which have potable water either private or unproductive.

The other important sub-basins with a number of hamlets distributed are the Thevaram, Andipatti and the Maruthanadi. Each has about 10 to 11 per cent of the hamlets of the Upper Vaigai basin. The most conspicuous feature in all these river basins are that hamlets with good drinking water facilities account for eighty per cent of the hamlets in the Maruthanadi, 50 per cent in the Thevaram and 29 per cent of the hamlets in the Andipatti sub-basins. As already mentioned most of these hamlets with good drinking water facilities are found in the lowland stretch of these rivers.

The next category of hamlets in the sub-basins is those belonging to the category having potable water but with non-perennial sources. They account for 62.8 per cent of the hamlets in the Andipatti sub-basin 38.8 per cent of the hamlets in the Thevaram basin and 22.6 per cent of the hamlets in the Suruliar basin. They are mostly found along the non-perennial streams draining the basin. The remaining hamlets in these basins belong to other different categories, their number being quite insignificant. The small group of sub-basins such as the Varahanadi, Koilpatti, Viruveedu and Kottagudi has 9.9 per cent, 5.9 per cent, 5.64 per cent and 5.56 per cent of the hamlets of the Upper Vaigai basin respectively. Of which 62 per cent of the hamlets in Varahanadi, 89 in Koilpatti, 32 per cent in the Kottagudi have hamlets with good drinking water sources.

The overall pattern of the distribution of hamlets with drinking water facilities in the Upper Vaigai basin are as follows: 53.1 per cent of the total hamlets in the region have good drinking water sources and these hamlets are distributed along the lower sections of the major tributaries draining the sub-basin. 27.6 per cent of the total number of hamlets in the study area has potable water from non-perennial sources and these hamlets are distributed mostly along the non-perennial streams of the tributaries flowing from the foot hills of the Varushanad region.

About 15 per cent of the hamlets in the study area have potable water with sources private or unproductive which is found distributed either in different parts of the low land section especially in the regions studded with tanks which are privately owned or in and around the estates in the Lower Palanis.

On the whole 95.3 per cent of the hamlets of the study area have the potable water and 4.4 per cent of the hamlets lack drinking water facilities. Of these 70 per cent has no source of potable water any where in the vicinity; 13.3 per cent has the source of water within the hamlet but water is not potable due to its impurities and 16.7 per cent of the hamlets have no potable water within a kilometer distance from the hamlet.

4.3.2.2 Urban Water Supply

In the Upper Vaigai basin 32.74 per cent of total population live in eleven urban centers of Theni-Allinagaram, Bodinayakkanur, Kambam, Periyakulam, Chinnamanur, Gudalur, Kodaikanal, Vattlagundu, Andipati-Jakkampatti and Highwavys. Table 4.5 and Figure 4.8 present urban drinking water supply per day. There are two sources available for the supply of urban drinking water viz., surface water sources and groundwater sources. Surface water is carried from the near-by river and groundwater is from the underground sources from wells. Almost all the towns in the basin receive surface water supply except Andipatti-Jakkampatti, which gets its water from the Vaigai river-bed. Periyakulam receives water both from surface and groundwater sources. Surface water is supplied from the Berijam Lake and the ground water is supplied from the wells in the Varahanadi bed. In other areas surface water being distributed for drinking purposes. Among these towns, Gudalur, Kambam, Uthamapalayam, Chinnamanur and Theni-Allinagaram get water from the Suruliar. Theni-Allinagaram is provided with 3919 m^3 of water daily. Kambam receives 2025 m^3 and Gudalur gets 1450 m^3 daily. Kambam receives 2025 m^3 and Gudalur gets 1450 m^3 daily. From the river Kottagudi, Bodinayakanur gets its daily water supply of 3810 m^3. Vattlagundu draws water from the Manjalar, which provides 308 m^3 daily. A total of 22648 m^3 of drinking water is being supplied daily to the urban centres in the basin. Water supply for these towns is not uniform throughout the year. The supply fluctuates according to the availability of water. In the dry months especially in summer most of the towns face acute water shortages.

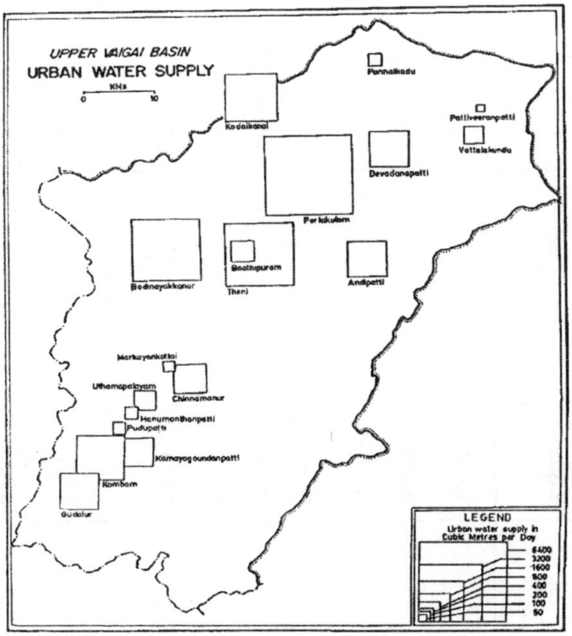

Fig. 4.8

Table 4.6 provides drinking water supply provided for other towns. A few schemes which are operating in the basin provide water supply to some of the villages.

The Suruliar water is supplied to the villages of Kamayagoundapatti, Pudupatti, Hanumanthanpatti and Markayankottai by the Kambam Valley Comprehensive Water Supply Scheme. Kamyagundenpatti gets 840 m^3 of water daily from the scheme. Other villages receive less than 200 m^3 of water per day. Devadanapatti, located in the Manjalar basin, receives water from the Manjalar Dam which provides 1330 m^3 water daily for its population. Pannaikadu gets water from the Maruthanadi River (146 m^3 of water) daily. From the Kottagudi River bed Boothipuram Village receives water.

4.3.2.3 Domestic Water Need

Based on the norms fixed by the Tamil Nadu Water Supply and Drainage Board the minimum domestic water need for rural population is 35 litres per day per head. For population in town

Table 4.5
Upper Vaigai Basin: Urban Water Supply

Urban Center	Population in 2001	Supply source		Quantity supplied		Total
		Surface water	Ground water	Surface water	Ground water	
Gudalur	36754	Suruliar	-	1450	-	1450
Kambam	50280	Suruliar	-	2025	-	2025
Uthamapalayam	19646	Suruliar	-	360	-	360
Chinnamanur	3796	Suruliar	-	871	-	871
Bodinayakkanur	58799	Kottagudi	-	3810	-	3810
Theni-Allinagaram	53023	Suruliar	-	3919	-	3919
Kodaikanal	20451	Pambar (Varahanadi tributary)	-	2275	-	2275
Periyakulam	44316	Berijam Lame	Varahan-adi bed	1230	5000	6230
Vattlagundu	18022	Manjalar	-	308	-	308
Andipatti-Jakkampatti	15396	-	Vaigai bed	-	1400	1400
Urban-Total	347483			16248	6400	22648

Table 4.6

Upper Vaigai Basin: Village Water Supply Schemes

Village	Population in 2001	Supply source		Quantity supplied		Total
		Surface water	Ground water	Surface Ground water water		
Kamayagoundampatti	12672	Suruliar	-	840	-	840
Pudupatti	8662	Suruliar	-	160	-	160
Hanumanthanpatti	7753	Suruliar	-	130	-	130
Markayankottai	5609	Suruliar	-	100	-	100
Boothipuram	7655	-	Kottagudi bed	-	420	420
Pannaikadu	7039	Maruthanadi	-	146	-	146
Devadanapatti	15630	Manjalar	-	1330	-	1330
Pattiveeranpatti	6439	-	maruthanadi bed	-	54	54

Panchayat, minimum requirement is 50 litres per day per head and for urban population 90 litres per head per day (Water Supply and Drainage Board, 1984).

Thus the Upper Vaigai basin requires 61,222.14 m^3 of water daily for its domestic purposes which works to 22.54 million m^3 per annum. Of the total requirements more than half (52.46) per cent of its requirement is needed for the urban population and the remaining 47.54 per cent by the rural population. Table 4.7 represents the domestic water need.

The total urban water need for the whole basin is 32, 1113.62 m^3 daily. Of which 41.44 per cent is the water need of the urban centers of the Suruliar basin followed by 31.33 per cent by the urban centers of the Kottagudi, 18.15 per cent by the Varahanadi, 5 per cent by the Manjalar and 4 per cent by the Andipatti sub-basins.

Only in the Suruliar, the Kottagudi and the Varahanadi sub-basins together the water need of the urban areas is more than 50 per cent of the domestic water need while in the Manjalar and the Andipatti basins it is around 40 per cent. The Thalai Vaigai, the Thevaram, the Koilpatti, the Maruthanadi, the Virualiaptti and the Viruveedu have only rural population.

As far as rural water need is concerned the Suruliar basin leads to the extent of 19.52 per cent followed by the Thalai Vaigai 16.72 per cent, the Thevaram 13040 per cent, the Varahanadi 11.72 per cent and the Manjalar 7.42 per cent.

From the above analysis it is observed that the water supplied for the urban centers are not uniform throughout the year. The supply fluctuates according to the rainfall and availability of water in the canals. An acute water shortage is noticed during the summer months. As far as villages are concerned most of the villages, with the exception of those along the river course face acute water shortages during the drier months. Thus there is an urgent need for comprehensive water resources planning for domestic purposes as well.

Summing up one could safely state that the total yield of water in the Upper Vaigai basin is 780.98 million m^3 comprising of 313.95

Table 4.7
Upper Vaigai Basin: Rural And Urban Water Need

Name of the sub-basin	Total		Rural		Urban	
	Drinking water need in m^3	in per cent	Drinking Water need in m^3	in per cent	Drinking water need in. m^3	in per cent
Thalai Vaigai	4923.72	8.04	4923.72	100.00	-	-
Suruliar	18896.23	30.86	5683.24	30.67	13212.99	69.23
Thevaram	3901.08	6.37	3901.08	100.00	-	-
Kottagudi	11587.22	18.92	1523.24	13.14	10063.98	86.76
Koilpatti	935.80	1.54	935.80	1000.00	-	-
Varahanadi	9240.07	15.09	3411.04	36.91	5829.03	63.09
Manjalar	3781.50	6.17	2159.52	57.10	1621.98	42.90
Maruthanadi	2941.60	4.80	2941.60	100.00	-	-
Viralipatti	374.32	0.63	374.32	100.00	-	-
Viruveedu	1242.96	2.04	1242.96	100.00	-	-
Andipatti	3397.64	5.54	2012.00	59.21	1385.64	40.79
Upper Vaigai Basin	61222.14	100	29108.52	47.54	32113.62	52.46

million m^3 (40.2 per cent) of surface water resources and 467.032 million m^3 (59.8 per cent) of groundwater resources. The quantity of water utilized for irrigation has been 397.75 million m^3. Of which 147.16 (37 per cent) was from surface water sources and 250.59 million m^3 (63 per cent) from ground water sources. There appears to be a surplus of water both from the surface and the ground after having met the requirements of agriculture. This surplus is to the tune of 383.2 million m^3 which amounts to 49 per cent of the total water yield of both surface and sub surface sources.

The demand for domestic water surplus is also quite high both from the villages and towns in the basin. The demand for rural domestic water is 10.6 million m^3 and for urban population is 11.7 million m^3 per annum.

The picture that emerges from these figures is optimistic as one finds a surplus of water resources – both from the surface and ground – to be around 383.2 million m^3 (49.06 per cent of the total yield) which would be enough for a long time to come at the present rate of utilization.

But in reality one finds that though the Upper Vaigai basin has ample water during and after monsoons, it has to supply water to the Lower Vaigai basin, which has an irrigated area of 56000 ha. A quantity of 17 million m^3 of water had been set apart for the release for the Madurai Corporation water supply. What evolves from this study is therefore better water management practices during and after the rains and conservation of the surplus water to be utilized in the dry months is essential.

CHAPTER 5

SUMMARY AND CONCLUSIONS

Water is a precious resource, indispensable for the existence and survival of life on the Earth. It is highly variable and mobile. As a result of increasing demand for water, the quantity of which is largely constant, there is a need for a detailed study to help modernize water use and seek measures to remove disparities and check imbalances while formulating water policies. An attempt has been made in this study to analyze the availability, development and utilization of water resources in the Upper Vaigai basin, Tamil Nadu. This sort of micro geographical study, of late, is assuming great importance in India.

According to the major objectives of the present study, the surface and ground water availability of the Upper Vaigai basin has been evaluated taking up each sub-basin separately and estimated the water balance. Analysis of the patterns of water resources development and utilization ha also been made. The study has used data from different sources and a number of techniques of water resource evaluation have been adopted.

The Upper Vaigai basin spreads over an area of about 3975 km², between 9° 31′ and 10° 20′ N latitudes and 77° 10′ and 77° 54′ E longitudes, in the Theni and Dindigul districts of Tamil Nadu. It is a part of one of the river basins of the State, the Vaigai basin. The basin is located in between the Lower Palanis, the Cardamom hills and the Varushanad - Andipatti hills of the Western Ghats and is drained by the Vaigai and a few tributaries of the Vaigai, the Suruliar, the Kottagudi, the Varahanadi, the Manjalar and the Maruthanadi. The Upper Vaigai basin has eleven sub-basins viz., the Thalai Vaigai the Suruliar, the Thevarm, the Kottagudi, the Koilpatti, the Varahanadi, the Manjalar, the Maruthanadi, the Viruveedu, the Viralipatti and the Andipatti.

The different geological formations broadly classified as the crystalline rocks of the Archaean era and the sedimentary formations of the quaternary era, constitute the geology of the study area. Various geomorphic zones such as hilly terrain, bajada, valley filled sediments, sand dunes, pediplain and flood plains can be identified in this region. The slope of the land varies from less than 1° to over 30°. The maximum slope is found along the crest of the Lower Palanis and the scrap edge of the Cardamom hills. The general direction of slope of the basin is form southwest and northeast.

The basin in general suffers from the semi-arid type of climate with the exception of the north and northwestern part, where climate is of per-humid type. Summers are hot when temperature rises to 36°C in May while the minimum temperature is 19.1°C in January. The average annual rainfall is around 880 mm, most of which falls during the monsoon seasons between June and November.

Tropical wet evergreen forests are found along the eastern slopes of the Cardamom hills and the western slopes of Varushanad hills; montane wet temperate forests are distributed in the Upper Kottagudi basin. The Lower Palanis are covered with sub tropical broad leaved hill forests while in the Agamalai, Kodaikanal, Varushanad - Andipatti hills, the tropical moist deciduous forests are common. Tropical dry deciduous forests are found in the eastern parts of the Lower Palanis while in the Andipatti hills shrubs and deciduous trees are the types of natural vegetation.

Three soil orders such as entisols, vertisols, and alfisols are generally found in the basin. Entisols, a recent soil, are red, non-calcareous, insitu, shallow and well drained. Vertisols, commonly referred to as black soil, are found in the center of the basin. Alfisols, reddish brown or red sandy soils, are found more extensively. Sand deposition and formation of san dunes are active in Kombai and adjacent areas of Bodinayakkanur.

The Upper Vaigai basin has a total number of 18016 streams in all segments. The river Vaigai is designated as an eighth order stream. The first order streams are more in the Thalai Vaigai and the Suruliar sub-basins. The bifurcation ratios range between 3.75 and 2.25 as the geological structures do not disturb the drainage pattern. Drainage density in the basin is 225 km/km^2.

The study area consists of 129 village and 11 towns and had a total population of 1,089,785 in 2001 with an overall density of 274 persons per square kilometer. Urban centers like Bodinayakkanur and Theni recorded high density of more than 2000 persons per km2. The urban population share is one-third of the total population of the basin.

A very uneven pattern of population distribution is noticed in the basin. The highest concentration of population is seen in areas along the Suruliar, the Kottagudi, the Varahanadi, and the Manjalar due to the perennial nature of the rivers and extensive low lands. The distribution, however, gets thinner in the Thevaram, the Andipatti, the Viruveedu and the Thalai Vaigai sub-basins due to the unduiating topography and non-perennial streams. The overall growth of population for the hundred year period 1901-2001 has been to the tune of 193 per cent. The growth was remarkably high between 1901 and 1911 largely due to the introduction of the Periyar Irrigation Project.

Land use pattern in the basin is quite unique in that forest covers an area of 37.1 per cent to the total area of the basin. Net sown area accounts for 36.2 per cent of the total area, of which 15 per cent is cropped more than once. The principal source of canal irrigation is the Periyar Irrigation Project. Canals as the largest source of irrigation are confined to the Suruliar, the Manjalar and the Maruthanadi sub-basins. Tanks and wells as sources of irrigation are found to exist in almost all parts of the basin.

Rainfall is the principal input in the hydrologic cycle, which influences the magnitude and time of runoff, soil moisture reserves and groundwater storage. The Upper Vaigai basin lies on the lee side of the Cardamom hills and so receives a low rainfall of around 880 mm per annum. The highlands of the Lower Palanis and the extreme southeastern part of the Varushanad hills receive a maximum of 1200 mm rainfall. The central part of the basin is a dry tract receiving less than 700 mm of rainfall per annum. The southwest and northeast monsoons together account for nearly 80 per cent of the total annual rainfall of the basin.

The regional trend of the annual rainfall is found to be sloping down from the west-northwest to east-southeast. In the west-northwest above 1000 mm of rainfall is expected each year and this declines to 700 mm in the east-southeast. The map of residuals from the trend surface shows that the highlands of the Lower Palanis and the Varushanad hills standout as areas with positive residuals due to high rainfall, whereas in the central stretch of the basin are the areas of negative residuals. The highlands, which receive high amounts of rainfall, have low rainfall variability compared to plains which receive low precipitation and suffers from high variability.

The potential evapotranspiration is very high, more than 1600 mm, in the central stretch of the basin. The moisture actually available for crops in the basin is less than 700 mm while the water need of the basin is around 1600 mm. Thus there is an acute water shortage in this area. Water deficiency is at its peak in the central stretch of the basin, where the water need is the maximum due to a comparatively dense population. Water deficiency ranges from 800 to 850 mm in other parts of the basin.

The total annual river flow varies from year to year. The percentage dependabilities of runoff have been worked out for Power House, Vaigai Dam, Peranai and Manjalar. The total annual river flow that has a return period of 20, 5 and 1.67 years of return period at Power House is 430, 520 and 640 million m^3 respectively, at Vaigai Dam it is 430, 590 and 840 million m^3 respectively and at Peranai it is 530, 643 and 860 million m^3 respectively. The surface flow of the basin is supplemented by a transfer of about 625 million m^3 of water from the neighboring Periyar basin.

During the dry summer months as could be expected the river flow is very low and the coefficient of variability is high. But for June to November, the monsoon period, the amount of river flow increases gradually with the variability decreasing. However, in the area of the Manjalar Dam there has been a greater stability in river flow during the northeast monsoons only, due to heavy rain and large runoff. The mean annual yield of the surface water in the Upper Vaigai basin is 314 million m^3. It is very high in the Varahanadi sub-basin which gets the benefits of heavy rainfall of the Lower Palanis and low in the Viralapatti sub-basin because of the limited areal extent coupled with low rainfall. The surface water of the basin is classified as good to permissible for irrigation, domestic and industrial supply.

The Odaipatti and parts of Seepalakottai micro basins of Thalai Vaigai sub-basin have more groundwater availability. The central part of the basin covering the Lower Suruliar have very deep groundwater table from where rises in all direction. In general, it is observed that the groundwater table is the deepest in the southwestern part of the basin and gradually rises towards the northeastern part.

The regional trend of mean annual groundwater table was found to be sloping down form the south-southwest to north-northeast. In the south-southwest zone 16 m depth in the groundwater table are expected each year and this rises to 6 m in the north-northeast. With regard to the groundwater table residual map the negative residuals coincide with very deep water table whereas the positive residuals coincide with areas where the groundwater table is close to the surface. The annual variability of water table is high in the south in the Varushanad Valley and in the northeastern part of the basin. The variability reaches the maximum during the northeast monsoon season.

A study of the trend of water table fluctuation during the period 1973-2002, reveals marked fluctuations spatially. There is a continuous trend of decline in the western part of the basin essentially due to the higher intensity of exploitation (pumping) whereas in the rest of the basin the water table fluctuates sharply within short distance of about a metre.

The total groundwater recharge is 550 million m^3. Of which 53 per cent is through irrigated fields and recharge through water bodies accounts the rest. The net recharge after allowing 15 per cent loss by evaporation is 467 million m^3. The total groundwater recharge through specific yield method is about 420 million m^3.

There is no significant change in the quality of groundwater in the study area. Groundwater tapped along river Vaigai locally shows slight chemical variations due to the solvent action in the weathered mantle and the induced infiltration from the river flow. A detailed survey conducted in each area revealed that the quality variation is a local phenomenon due to the entry of domestic and municipal sewerage. The suitability of groundwater for irrigation in the basin is classified as good to permissible.

The agricultural prosperity of the Upper Vaigai basin has considerably improved after the completion of the Periyar Irrigation Project. The major irrigation projects such as Periyar Irrigation Project and Vaigai Dam irrigate about 10 per cent of the total irrigated area of the basin. Of the total area irrigated in the basin well irrigation contributed about 70 per cent.

Summing up one could safely conclude that the total yield of water in the Upper Vaigai basin is 781 million m^3 comprising 40.2 per cent (314 million m^3) of surface water and 59.8 per cent (467 million m^3) of groundwater resources. The quantity of water utilized for irrigation is 397 million m^3 of the total water resources, of which 37 per cent (147 million m^3) are from surface water sources. Thus it appears that groundwater resources are exploited mercilessly. There appears to be a surplus of water both from the surface and groundwater after having met the requirements. This surplus is to the tune of 383 million m^3 i.e. 49 per cent of the total water resources of the basin. But in reality one finds that though the Upper Vaigai basin has ample water during and after monsoons it is committed to supply water to the Lower Vaigai basin, which has an irrigated area of 56000 ha.

It is observed that more than 50 per cent of the settlements have good sources of drinking water supply throughout the year while 28 per cent of the settlements have non-perennial sources of drinking water and face a shortage during the dry seasons from March to May. About 15 per cent of the hamlets have potable water with sources either private or unproductive and the rest lacks drinking water facilities.

The demand for domestic water supply is quite high from the settlements of the basin, both rural and urban. The demand for rural domestic water was 10.6 million m^3 and for urban population 11.7 million m^3 per annum.

As the demand for water grows both in relation to the absolute requirements and with regard to variety of uses it is put to, unmanaged water resources will become less able to supply the needs. Therefore care should be taken for better water management practices during and after the rains and conservation of surplus water to be utilized in the dry months. It is also important that ground water exploitation is minimized both for irrigation and domestic supply.

From the above it is clear that the Upper Vaigai basin calls for efficient management of water resources. This study has succeeded in assembling and presenting facts about water resources, its use and development programmes in the basin. It has limited aims. This study when extended can lead to a proper management programme of water resources. The study has been fairly successful in understanding the precipitation, river flow, and water quality characteristics. However, the period of data available for groundwater table is limited and geophysical investigations are limited to certain areas only. The accuracy of yields of water will be sharp when data is available for longer periods.

In-depth studies of the Upper Vaigai basin can be made on the following lines: Rainfall characteristics and crop growth can be monitored simultaneously at short intervals of time. This will furnish an idea of water requirements of different crops. The above study can also lead to an understanding of the loss of water due to over irrigation.

The following measures should be taken to mitigate the problems faced by the Upper Vaigai basin. The feasibility of transfer of surplus water from the sub-basins of the Suruliar, the Kottagudi and the Varahanadi to the water deficient areas such as the sub-basins of the Thevaram and Andipatti can be studied in detail. Exploitation of groundwater should be minimized in the sub-basins of the study area.

The Periyar Lake, the Vaigai dam, the Manjalar dam and the Maruthanadi dam have experienced an acute problem due to silting thus reducing considerably the quantity of water in these areas. Desilting operations should be carried on at regular intervals to enhance the water supply to the irrigated lands. Seepage of water from canals is more in the basin. Considerable amount of seepage can be reduced by the proper lining of canals. The choice of crops especially in the sub-basins with a high water deficiency should be the drought resistant varieties. Pollution of surface and groundwater by urban sewerage disposal and disposal of other effluents should be avoided or at least minimized by proper treatment before they are let into the streams. This will help in maintaining the quality of water. It is essential that conjunctive utilization of surface and groundwater resources should be planned for irrigation commands. This will help to irrigate much larger areas, even with the existing water resources.

Appendix - 1
Upper Vaigai Basin: Climatic Parameters

	Jan	Feb	Mar	Apr	May	June	July	Aug	Sept	Oct	Nov	Dec	Mean Monthly
Mean Maximum Temperature in °C													
Kodaikanal	17.0	17.9	19.2	19.8	20.4	18.5	17.1	17.5	17.7	17.0	16.1	16.4	17.9
Kamatchipuram	31.2	32.7	35.0	36.0	36.0	33.9	34.0	33.9	33.7	31.9	31.2	30.5	33.3
Viralipatti	34.8	36.2	38.1	39.6	38.8	37.1	36.4	35.9	35.6	34.7	32.9	31.6	36.0
Madurai	29.5	32.2	34.5	36.6	37.0	36.3	35.5	34.5	33.7	32.5	3.8	28.7	33.5
Mean Minimum Temperature in °C													
Kodaikanal	8.0	8.7	10.0	11.6	12.7	12.2	11.5	11.5	11.3	10.9	9.8	8.6	10.3
Kamatchipuram	15.1	15.7	18.5	20.3	21.3	20.4	20.7	20.4	20.1	20.4	19.3	17.0	19.1
Viralipatti	17.0	17.4	20.3	22.4	23.0	22.5	22.2	21.7	21.2	22.3	20.5	18.0	20.7
Madurai	20.0	20.9	22.3	24.8	27.8	26.0	25.2	24.9	24.0	23.5	22.7	21.1	23.6
Relative Humidity in per cent													
Kodaikanal	69.0	63.5	58.5	70.5	74.5	81.5	85.0	85.0	84.5	88.5	87.5	76.0	77.0
Kamatchipuram	74.2	67.6	66.6	68.8	65.1	67.1	68.4	69.1	69.3	76.5	72.3	70.7	69.7
Viralipatti	58.3	53.8	54.5	56.1	55.4	56.6	56.4	57.5	60.6	67.5	72.0	66.1	59.6
Madurai	82.4	84.1	82.4	78.3	72.4	69.1	67.9	70.4	75.5	80.6	81.7	80.4	77.1

Sunshine in hrs													
Kodaikanal	6.1	7.1	6.6	5.8	5.0	4.9	5.2	4.2	7.6	5.0	5.4	5.7	5.7
Kamatchipuram	-	-	-	-	-	-	-	-	-	-	-	-	-
Viralipatti	9.0	9.3	9.1	8.6	7.9	5.6	4.6	5.6	6.5	6.2	5.5	6.9	7.1
Madurai	10.0	10.7	10.0	9.5	8.9	7.6	7.0	7.8	8.2	7.1	6.6	7.7	8.4
Wind Velocity in Kmph													
Kodaikanal	13.1	12.1	13.5	12.4	11.8	14.6	16.7	13.6	11.5	10.1	12.1	12.7	12.9
Kamatchipuram	-	-	-	-	-	-	-	-	-	-	-	-	-
Viralipatti	7.9	7.1	6.3	10.7	10.7	15.0	14.3	15.0	10.7	6.9	6.0	9.2	9.7
Madurai	8.6	8.8	8.1	10.0	10.0	12.1	10.3	8.8	5.8	8.4	7.9	7.4	8.7

Appendix - 2
Upper Vaigai Basin: Soil

Soil series/order	Parent Material	Characteristics	Slope in degrees	Erosion	Water Table in m	Drainage and permeability
Palaviduthi - Alfisols	Colluvium	Yellowish red to dark red, very deep sandy loam, non calcareous, colluvial soil and slightly acid	1-8	Slight to severe	10-20	Moderate to well drained and moderately permeable
Vylogam - Alfisols	Weathered gneiss	Yellowish red to dark red, very deep sandy loam, non calcareous, lateritic soil and slightly acid to neutral	1-3	Slight to moderate	7-20	Well drained and moderately permeable
Palathurai - Alfisols	Weathered gneiss	Red to dark brown, moderate to deep sandy loam, slightly gravel, clay loam calcareous, insitu, neutral to moderately alkaline	1-3	Slight to severe	10-40	Well drained and moderately permeable
Thulukkanur - Alfisols	Weathered gneiss	Dark Yellow brown, moderately deep, clay loam, sandy clay loam, calcareous insitu and mild to strongly alkaline	1-3	Slight to moderate	7-15	Moderate drained and moderately permeable
Somayanur - Alfisols	Colluvium	Reddish brown, very deep, colluvial soil, sandy loam to sandy clay loam, calcarious and slightly acid to neutral	1-3	Slight to moderate	12-20	Well drained and moderately permeable
Irugur - Entisols	Weathered gneiss	Yellowish red to dark red, shallow to moderately deep, sandy loam, non calcareous, insitu, soil, and slightly acid to neutral.	1-5	Slight to severe	10-20	Well drained and moderately permeable
Pilamedu - Vertisols	Weathered gneiss	Grey to very dark grey, deep to very deep, calcareous, sandy loam, sandy clay loam, and mild to strongly alkaline	1-3	Slight to moderate	20-40	Moderate drained and moderately permeable
Alathur - Vertisols	Alluvium	Dark grayish brown to very dark grayish brown, very deep, clay loam, slightly calcareous with lime concretion at lower depth, and neutral to mild alkaline	1-3	Slight	10	Moderate drained and moderately permeable

REFERENCES

Abbi, S.D.S., D.K. Gupta and S.K. Subramanian, (1970). On Some Hydrometeorological Aspect of Narmada Basin, ***Indian Journal of Meteorology and Geophysics,*** 21 (4), pp. 539-543.

Abhu, Lakshmi Singh and Lalita Salvia, (1993). Utilization of Water Resources for Agricultural purposes in Western Uttar Pradesh. ***Journal of Regional Science***. 25(1), pp. 49-58

Ahmed, M.R., (1976). The Highest Head Hydro-Electric Project - Suruliar in Tamil Nadu, ***Indian Journal of Power and River Valley Development,*** January, 1976, pp.1-9.

Ahmed, M.R., (1977). The Fortune of Madurai District in Tamil Nadu - An Ancient Irrigation Project converted into a Power Project, ***Irrigation and Power Journal,*** January. 1977, pp. 39-45.

Akiyama, M., (1980). The Recent Changes of Water uses in the Lower Reaches of the Takahoshi River, Western Japan, ***Geographical Review of Japan,*** 53 (11), pp. 679-698. (in Japanese and abstract in English).

Ali, S.H., (1980). Water Management in Command Areas of Andhra Pradesh, Institution of Engineers (India), ***Bulletin,*** 30, pp. 12-13.

Anandaraj, M., (1978). ***Water Balance studies in Tambraparni Basin***. Madras: College of Engineering (Mimeographed).

Anandarao, (1958). ***Development of Irrigation in Madras State***. Madras: Public Works Department.

Arunachalam, B., (1983). Water Resources of India in Indian Council of Social Science Research (1983). ***A Survey of Research in Physical Geography***. New Delhi: Concept Publishing Company.

Aston, A.R. and F.X. Dunin, (1980). Land-Use Hydrology: Shoalhavon, New South Wales, ***Journal of Hydrology***, 48, pp. 71-81.

Ayoade, J.O., (1975). Water Resources and their Development in Nigeria, ***Hydrological Sciences Bulletin***, 20 (4), pp. 581-591.

Ayangar, P.S.R., (1932). The Kambam Valley, ***Journal of the Madras Geographical Association***, 7 (3), pp. 278-286.

Baliga, B.S., (1960). ***Madras District Gazetteers : Madurai.*** Madras: Government of Madras.

Baker, M.N. and R.E. Horton, (1936). Historical Development of Ideas regarding the Origin of Springs and Ground Water, ***Transactions of the American Geophysical Union***, 17, pp. 395-400.

Balakrishna, S. and D.C. Reddy, (1978). Progress of Rural Water Supply Scheme under the Minimum – Needs Programme in Andhra Pradesh, ***Behavioral Sciences and Rural Development***, 1 (1), pp. 25-32.

Balakrishnan, G., R. Sakthivadivel and S. Nagarathinam, (1972). An Investigation of Ground Water Potential in Tamil Nadu, in proceedings of the ***Seminar of Problems in Ground Water Development***. Madras: Institution of Engineers (India), pp. 18-24.

Bannerjee, K.C., (1944). Regulation of River Flood: The Problem of Damodar River, ***Calcutta Geographical Review***, 5 (3), pp. 107-110.

Bansil, P. C., (2004). ***Water Management in India.*** New Delhi: Concept Publishing Company

Batisse, M., (1976). Water Needs and Resources: Rational Management: the Key to the Supply – Demand Equation, ***Nature and Resources***, 12 (4), pp. 2-9.

Bekinsale, R.P., (1969). Human Responses to River Regimes in Chorley, R.J., (1969). ***Water, Earth, and Man***. London: Methuen and Co. Ltd.

Bhati, Vinita M., (1996). ***Water Resources and their Utilization.*** Jodhpur: Vigyan Prakashan.

Bilas, R., (1980). Ground Water Resource of Varanashi District, India: An Assessment of Condition, Use and Quality. ***The National Geographical Journal of India***, 26 (1-2), pp. 81-93

Biswas, A.K., (1969). A short history of Hydrology, ***The Progress of Hydrology***, 2. pp. 914-936.

Biswas, A.K., (1970). ***History of Hydrology***: The Progress in Hydrology. Urbane: University of Illinois.

Bohle, H., (1976). Problems of Utilization of Water Resources in the Cauvery Delta – A case study of Talanayar Block in All India ***Symposium on Resource Development and Planning***. Madras: University of Madras.

Bose, S.C., (1948). ***The Damodar Valley Project***. Calcutta: Phoenix.

Bouwer, H., (1978). ***Ground Water Hydrology***. New York: McGraw Hill Book Company.

Bowden, D.J., (1980). Rainfall in Sierra Leone, ***Singapore Journal of Tropical Geography,*** 1 (2), pp. 31-39.

Bramson, R.L., P.F. ratt, J.D. Rhoades and J.D. Oster, (1975). Water Quality in Irrigated Watersheds, ***Journal of Environmental Quality,*** 4, pp. 33-40.

Brassel, K.E., and D. Reit, (1979). A Procedure to generate Thiessen Polygon, ***Geographical Analysis,*** 11 (3), pp. 289-303.

Brutsaert, W.F., and T.G. Gebhard, (1975). Conjunctive Availability of Surface and Ground Water in the Albu Guergne Area, New Mexico, ***Ground Water,*** 13 (4).

Central Ground Water Board, (1980). ***Ground Water Resources of Tamil Nadu***. Hyderabad: Central Ground Water Board, Southern Region, (Mimeographed).

Central Ground Water Board, (1983). ***Ground Water Resources of Noyil, Ponnani and Vattamalaikarai Basins***. Hyderabad. Central Ground Water Board (Mimeographed).

Chakravorthi, A.K., (1959). Floods in Himalaya Rivers of Ganga Plain. ***Geographical Review of India,*** 21 (2-4), pp. 75-81.

Chandrahasan, K., (1978). The Climate of the Palani Hills, ***The Indian Geographical Journal,*** 53 (2), pp. 51-58.

Chandrasekharan, H., R. K. Sharma, and K. V. Sundaram, (Ed) (2004). ***Water Resources Development and Management***. New Delhi: Mital Publication.

Chang, J., (1982). Stream Order and Stream Net Ratios of Drainage Basins in Taiwan, ***Geographical Research,*** 8, pp. 113-156. (In Chinese and abstract in English).

Chorley, R.J. and R.W. Kates, (Ed.) (1969). ***Water, Earth and Man.*** London: Methuen.

Chatterjee, B.N. and K. Bagchi, (1961). Floods in West Bengal, ***Geographical Review of India,*** 23 (3), pp. 125-131.

Chaturvedi, A.C., (1977).Planning for Ground Water Resources, ***Journal of the Institution of Engineers (India)***, 58 pp. 103-105.

Chaturvedi, M.C., (1976). ***Second India Studies Water.*** Delhi: The Macmillan Company of India Ltd.,

Chidley, T.R.E. and K.M. Keys, (1970). A Rapid Method of Computing Areal Rainfall, ***Journal of Hydrology***, 12, pp. 15-24.

Chinnamani, S. and R. Sakthivadivel, (1981). ***An Integrated Study of Hydrology of the Bhavani Basin, Part I – Natural Resources,*** Madras: Perarignar Anna University of Technology (Mimeographed).

Chow, V.T., (1964). ***Hand Book of Applied Hydrology***. New York McGraw-Hill Book Company.

Daniel, J.K.K., (1981). Drainage Density as an Index of Climatic Geomorphology, ***Journal of Hydrology***.50, pp. 147-154.

Darcy, H., (1856). ***Les Fontaines Publiques de la Vile de Dijon***. Paris: Victor Dalmont.

Datta, P.S., B.I. Desai and S.K.Gupta, (1979). ***Comparative Study of Ground Water Recharge Rates in Parts of Indo-Gangetic and Sabarmati Alluvial Plains,*** Mausam, 30 (1), pp. 129-133.

Davis, S.N. and R.J.M. Dewiest, (1967). ***Hydrology.*** New York: John Wiley & Sons Inc.

Dayal, P., (1971). River Valley Development Planning: A trend Report in ICSSR (1971). ***A Survey of Research in Geography.*** New Delhi. ICSSR, pp.134-144.

Dawbree, A., (1887). ***Les Eaux Souterraines anx Epoques Anciennes at a 1' epoque actuelle.*** Paris: Denold.

Dey, B., (1978). The Water Balance and Its Role in West Bengal's Agriculture, ***Indian Journal of Regional Sciences***, 1 (2), pp. 10-17.

Douglas, I., (1972). The Geographical Interpretation of River Water Quality Data in C. Board, Peter Haggett, and David r. Stoddart (Eds.) (1972). ***Progress in geography: International Review of Current Research***. 4. London: Edward Arnold.

Dutt, P.K., (1948). The Ganges Canal Hydro-electric Scheme and Aid to Agriculture in Western U.P., ***The Indian Geographical Journal,*** 23 (4), pp. 12-23.

Eagleson, P.S., (1970). ***Dynamic Hydrology.*** New York: McGrew-Hill Book Company.

Efremov, J.K., (1961). Nature Conservation Law in R.S.F.S.R. and Tasks of the Soviet Geographers, ***Vestiva Geogr. Obschestra***, 93, pp. 193-198.

El-tom, M.A., (1969). A Statistical Analysis of the Rainfall over the Sudan, ***The Geographical Journal***, 135 (3), pp. 378-387.

Furon, R., (1967). ***The Problem of Water: A World Survey.*** London: Faber and Faber Ltd.

Gandhi, P.R., (1978). Distribution of Water Resources and Water Management in the Drought-Prone Areas in the Maharastra State, ***Irrigation and Power Journal***, 35 (4), pp. 445-456.

Ganesh, A., (1984). Water Resources Development in Lower Mekong Basin, ***The Geography Teacher (India)***, 29 (5), pp. 6-16.

Gokhale, B.K., (1959). The Gungabhadra Project, ***Indian Journal of Power and River Valley Development***, 9 (2), pp. 54-66.

Gosain, A. K and Sandhya Rao, (2004) GIS - based technologies for Watershed Management, ***Current Science***. 87, pp. 948-953.

Goswami, A.C., (1972). Regional Flood-Frequency Analysis of Brahmaputra Basin in North East India, ***Journal of the Institution of Engineers (India)***, 52 (7), C 14, pp. 169-174.

Government of Tamil Nadu (1974). ***Water Atlas of Tamil Nadu.*** Madras: Ground Water Division.

Government of Tamil Nadu, (1978). ***Survey on Drinking Water Sources of the Rural Areas.*** Madras: Government of Tamil Nadu (Mimeographed).

Greer, C., (1979). ***Water Management in the Yellow River Basin of Chin.*** Austin: University of Texas.

Gregory, K.J. and D.E. Walling, (1976). ***Drainage Basin Form and Process: A Geomorphological Approach.*** London: Edward Arnold.

Gupta, S. K., and R.D. Deshpande (2004). Water for India in 2050: First Order Assessment of Available Options. ***Current Science***. 89, pp. 1216-1224

Gumbel, E.J., (1954). ***Statistical Theory of Extreme Values and some Practical Applications.*** US Department of Commerce, National Bureau of Standards.

Haragan, D.R., (1976).Spatial Variation of Precipitation on the Texas High Plains, ***Water Resource Bulletin***, 12 (6), pp. 216-221.

Hart, H.C. (1956). *New India Rivers.* Calcutta: Longmans.

Hart, W.C., (1932).The Forests of Madura District, ***The Journal of the Madras Geographical Association,*** 7 (3), pp. 196-200.

Hemamalini, B., (1992). Water Balance Analysis, ***Annals of the National Association of Geographers, India.*** 12 (1&2), pp. 15-18.

Hidore, J.J., (1970). Water, Water Resources, and Water Problem, ***The Professional Geographer,*** 22 (3) pp. 117-119.

Hirose, T. and N. Kuramoto, (1981). Stream Water Quality as influenced by Land Use Patterns in the Kakioka basin, Japan, ***Journal of Environmental Quality,*** 10 (2), pp. 184-188.

Hjelmfelt, A.T., (1978). Amazon Basin Hydrometeorology, ***Journal of the Hydraulic Division,*** ASCE, 104 (6), pp. 887-897.

Hollis, J. and J. McEvoy III, (1973). Demographic Effects of Water Development in Goldman, Charles R.J. McEvoy III and P.J. Richarson, (Eds.) (1973). ***Environmental Quality and Water Development***. San Francisco: W.H. Freeman and Company.

Horton, R.E., (1945). Eros ional Development of Streams and their Drainage Basins: Hydro Physical Approach to Quantitative morphology, ***Geological Society of American Bulletin,*** 56, pp. 275-370.

Institute for Water Studies, (1983). ***Annual Report of the Institute for Water Studies.*** Madras: Institute for Water Studies.

Institute of Hydrology, (1971). ***Research 1970-71,*** NERC: Institute of Hydrology.

International Institute for Land Reclamation and Improvement (1974), ***Drainage Principles and Applications, III Surveys and Investigations,*** Wageningen, The Netherlands: ILRI.

Irrigation Commission, (1972). ***Report of the Irrigation Commission***. New Delhi: Government of India.

Iyer, A.S. and B.M. Singh, (1981). Use of Aerial Photo Interpretation technique for Watershed Management Studies in Draught Prone Areas, ***Journal of Photo-Interpretation and Remote Sensing,*** 9 (1), pp. 15-19.

Jain, P.K., (1968). Development of Water Resources of Western Rivers of Uttarpradesh, ***Indian Journal of Power and River Valley Development,*** 17 (12). Pp. 312-316.

Jauhari, A.S., (1956). Development of Electric Power in North West India, ***National Geographical Journal of India***, 2 (1) pp. 48-67.

Jeyapalan, B.A., (1979). Water Management of Tank Irrigation System in Tamil Nadu, ***The New Irrigation Era***, 17 (1), pp. 12-15.

John, G. Lyon (2003). ***GIS for Water Resources and Watershed Management***. London and New York: Taylor and Francis

Johnson, B.L., (1969). Lower Mekong Basin: A Review, ***Geographical Journal***, 135 (4) pp. 588-613

Johnson, D.M. and J.O. Dart, (1982). Variability of Precipitation in Pacific North West: Spatial and Temporal Characteristics, ***Water Resources Research Institute Publications***, 2, pp. 327-333.

Jones, P.B., G.D. Walker, R.W. Harden, and L.L. McDaniels, (1963). ***The Development of the Science of Hydrology Texas***: Texas Water Commission.

Kalpalatha, S., (1983). ***Watershed Management in Chinnar Reservoir Catchment***. Unpublished Ph.D. Thesis. Madras: University of Madras.

Kazmann, R.G., (1965). ***Modern Hydrology***. New York: Harper and Row.

Keller, R., (1968). The Role of Geography with in the International Hydrological Decade, in R. Keller (ed.) (1968). ***Runoff Regiments and Studies of the Water Balance.*** Feiburg, F.R. Germany: Selbstvelag Der Geographichen, Institute Der Alber-Ludwig Universitate.

Khosla, A.N., (1951). ***Appraisal of Water Resources: Analysis and Utilization of Data in Proceedings of the Resources***, 4 Water Resources. New York: United Nations Department of Economic Affairs.

Kirk, W., (1950). The Damodar Valley - Vales Opima, ***Geographical Review***, 40, pp. 415-443.

Klimaszewski, M., (1956). The Detailed Hydrographical Map of Poland, ***Przegl. Geogr***. 28, Supple, 4147.

Kos, Z., (1981). Conjunctive Use of Surface and Ground Water in the Istra Peninusla, Yugoslavia, ***Water International***, 6 pp. 122-125.

Krishnamurthy, N., (1940). Papanasam: Hydroelectric Project, ***The Journal of the Madras Geographical Association***, 15 (2), pp. 112-124.

Kulkarni, V.M., Dehai and S.M. Dorges, (1968). Amba River: Drainage Basin Characteristics, ***Bombay Geographical Magazine***, 14 (1), pp. 35-38.

Kumaraswamy, P., (1974). ***Surface Water Availability in Tamil Nadu.*** Poondy: Institute of Hydraulics and Hydrology (Mimeographed).

Kumaraswamy, P., (1978). ***Mathematical Models of River Basins in Tamil Nadu Part 2 Mathematical Models.*** Poondi: Institute of Hydraulics and Hydrology.

Leszczycki, S., (1960). The Application of Geography in Poland, ***The Geographical Journal,*** 126 (4), pp. 418-426.

Linsley, R.K. and J.B. Frangini, (1972). ***Water Resources Engineering.*** Tokyo: McGraw-Hill Book Company.

Lloyd J.W., D.S.H. Drennan and B.M.U. Bennel (1966). A Ground Water Recharge Study in North East Jordan, ***Proceedings of the Institute of Civil Engineers,*** 35, pp. 615-631.

Lvovich, M.I., (1959). ***The Water Balance of the Land,*** Materialik IIIs, zdu. Geogr. Obsch. U.S.S.R.

Lvovich, M.I., (1959). The Water Aspects of Geography of Agriculture in USSR, ***Soviet Geographer,*** A 20 (9) pp. 515-521.

Narasimhan, T. N., (2005) ***Water: Science and Society,*** 89, pp. 787-793.

Madhuramma, Ch. and M. S. Rao, (1993). Water Balance and Cropping Pattern of Nellore District, A.P., ***National Geographer,*** 28 (1), pp. 1-19.

Mathur, R.N., (1969). ***A Study in the Groundwater Hydrology of the Meerut District, Uttar Pradesh, India,*** Varanashi: Banaras Hindu University.

Mathur, R.N. and K.S. Yadav, (1980). Assessment and Management of Water Resources: A Case Study of Schchabyanj Block, Chakia Tabsill, Varanasi, ***Transaction in the Institute of Indian Geographers,*** 2 (2), pp. 101-105.

Meinzer, O.E., (1923). The Occurrence of Groundwater in the United States. ***UGGS Water Supply Paper,*** 489.

Michael, A.M., (1978). ***Irrigation: Theory and Practice.*** New Delhi: Vikas Publishing House Pvt. Ltd.

Monkhouse, F.J. and H.R. Wilkinson, (1967). ***Maps and Diagrams: Their Compilation and Construction.*** London: Methuen and Co.,

Mostertman, L.J., (1977). Water Quality: An Overview, ***Agri-Eco Systems,*** 3. pp. 239-251.

Murty, G.S., (1971). ***Geology of Parts of Uthamapalayam and Usilampatti Taluks, Madurai District, Tamil Nadu*** Madras: Geological Survey of India, Tamil Nadu Circle (Mimeographed).

Murty, G.S., (1975). ***Geology of Parts of Uthamapalayam and Usilampatti Taluks, Madurai District, Tamil Nadu*** Madras: Geological Survey of India, Tamil Nadu Circle (Mimeographed).

Muthuswamy, T.N., (1932). The Physical Geography of the Madura District, ***The Journal of the Madras Geographical Association,*** 7 (3), pp. 183-191.

Muthuswami A.P. (1983). ***Total population and Population of Scheduled Castes and Scheduled Tribes in Panchayats and Panchayat Unions: Madurai District.*** Madras, Government of Tamil Nadu.

Naegamvala, J.P., (1961). Narmada River Project, ***Indian Journal of Power and River Valley Development,*** 11 (6pp. 1-8.

Nag, B.S. and G.N. Kathpalia, (1975). Water Resources of India in IWRA, (1975). ***Water for Human Needs,*** 2, pp. 373-382, New Delhi: IWRA.

Narayan, M.V., (1976). ***Prediction of Runoff for Sub-basin of the Upper Cauvery Catchments.*** Mysore: National Institute of Engineering.

National Remote Sensing Agency, (1979). ***Satellite Remote Sensing Survey: Southern Part of Tamil Nadu.*** Secunderabad: NRSA (Mimeographed).

Nejand, S. and R.A. Khan, (1975). Ground Water Quality in the Shirag Basin, South Iran, ***Ground Water,*** 13 (3), pp.

Niyogi, M. and K. Bagchi, (1981). Quality of Ground Water in the Ajoy Basin, Indian ***Journal of Power and River Valley Development,*** March-April, 1981. pp. 46-51.

Olson, R.E., (1970). ***Geography of Water.*** IOWA: W.M.C. Brown Company Publishers.

Oppen, M.V. and H.P. Binswanger, (1977). Institutional and Physical Factors Affecting Tank Irrigation Density in India in the ***Seventeenth Indian Econometric Conference,*** Trivandrum, Dec. 1977, Ulloor: Centre for Development Studies.

Oppen, M.V. and K.V.S Rao, (1980). ***Tank Irrigation in Semi-Arid Tropical India. Part I: Historical Development and Spatial Distribution.*** Patancher: International Crops Research Institute for the Semi-Arid Tropics (ICRISAT).

Oskolkova, N.D., (1978). Estimation of the Change of the Ground Water Regimein the Chu River Basin under the Effect of Envirrnmental and Reclamation Factors, ***Soviet Hydrology,*** 18 (4), pp. 315.321.

Patel, A. S., (2004). Water Resources Management and People's Participation. ***Journal of Applied Hydrology,*** pp. 1-7

Padmanabhamurthy, B. and B.C. Biswas (1970). A Comparative Study of Potential Evapotranspiration by Different Methods, ***Indian Journal of meteorology and Geophysics,*** 22, pp. 238-240.

Panchanathan, S., (1976), Ground Water Evaluation Methods and Techniques in ***Symposium on Resource Development and Planning,*** November, 1976, Madras: University of Madras.

Pathak, B.D., (1978). Exploration and Exploitation of Ground Water with Special Reference to Recharge, ***The National Geographical Journal of India,*** 24, (3 and 4), pp. 9-54.

Penman, H.L., (1948). Natural Evaporation from Open Water, Bare Soil and Gras, ***Proceedings of the Royal Society London, A,*** 193, pp. 120-146.

Planning Commission, (1978). ***Report of the Planning Commission.*** New Delhi: Government of India.

Pleva, E.G., (1961). Multipurpose Land and Water Districts in Ontario in J. Hendry, (Ed.) (1961), ***Comparisons in Resource Management.*** Baltimore: John Hopkins Press.

Public Works Department, (1974). ***Conference on Madurai District Irrigation.*** Madras: PWD. (in Tamil).

Public Works Department, (1976). ***Report on the Ground Water Studies in Vaigai Basin.*** Madurai: Ground Water Circle (Mimeographed).

Rai, R.K., and B. Dutta, (1982). Geomorphology and Ground Water Potential of Garo Hills, Meghalaya District, ***Hill Geographer,*** 1 (1), pp.17-23.

Rajan, B.H., (1978). Optimum Use of Irrigation Water, ***Indian Journal of Power and River Valley Development,*** June, 1978, pp.155-159.

Rakhecha, P., (1974). The Study of Draught by Water Budget Method over A.P., ***Indian Journal of Meteorology and Geophysics,*** 24 (3 and 4), pp.411-416.

Ramamoorthy, M., (1963). Power Development in South India, ***Indian Journal of Power and River Valley Development,*** 13 (9), pp.13-18.

Ramesh, A., (Ed.), (1984). ***Resource Geography***. New Delhi: Heritage Publishers.

Ramaniah, Y. and K. Ravindra Reddy, (1997). A Surface Water Balance Study for Chitravathi River Basin, Andhra Pradesh, ***The Indian Geographical Journal,*** 70 (1), pp. 34-42.

Rammohan, H.S., (1978). ***A Study of the Water Balance and Draught Climatology of Tamil Nadu,*** Unpublished Ph.D. Thesis. Waltair: Andhra University.

Rao, C.S., (1978). The Effect of Water Table Fluctuation on Water Quality: A Case History ***Indian Journal of Meteorology, Hydrology and Geophysics,*** 29 (4) pp. 691-694.

Rao, K.L., (1975). ***India's Water Wealth: Its Assessment, Uses and Projections***. New Delhi: Orient Longman Ltd.

Rao, K.N., J. George and K.S. Ramasastri, (1976). The Climatic Water Balance of India, ***Memoirs of the India Meteorological Department,*** 23 (3), pp. 42-47.

Rao, K.V.S., M. Ganesan and D.P. Ghose, (1971). Techniques and Methodology for Assessing Surface Water Resources of a River Basin, in ***Symposium on Integrated Development of Surface and Sub-surface Water Resources,*** New Delhi: Irrigation and Power Commission.

Rao, K.L., (1968). Water Resources of Tamil Nadu, in A.R. Irawathy (Ed.) (1968). ***This is Tamil Nadu.*** Madras: IGU Regional Congress, pp. 26-36.

Rao, S.S., and V.P. Subrahmanyam, (1968). Estimating of Yields from River Basins by a Modification of the Water Balance Procedure of Thornthwaite, ***Indian Journal of Meteorology and Geophysics,*** 12 (1-4), pp. 339-345.

Rathor, H.S. and S.N. Singh, (1970). Climatic Water Balance for Uttar Pradesh and Vicinity, ***Geographical Review of India,*** 33 (5), pp. 157-168.

Ray, K.B., (1950). River Valley Projects, ***Calcutta Geographical Review,*** 12 (3), pp. 51-60.

Reddy, G.P. Obi, K. Chandramouli, S. K. Srivastav, C.V. Srinivas, and A. K. Maji, (2000). Evaluation of Ground Water Potential Zones Using Remote Sensing Data - A Case Study of Gaimukh Watershed, Bhandara District Maharashtra, ***Journal of the India Society of Remote Sensing,*** 28 (1), pp. 19-32

Reddy, K.R., and N.B.K. Reddy, (1983). Surface Water Resources in the Swarnamukhi Basin. ***National Geographer***, 18, pp.19-28.

Rees, J.A., (1973). The Demand for Water in South-east England, ***The Geographical Journal***, 139 (1), pp. 20-42.

Remson, I. and J.R. Randolph, (1958). Application of Statistical Methods to the Analysis of Ground Water Levels, ***Transactions of American Geophysical Union***, 39 (1), pp. 75-83.

Richards, L.A., (Ed.) (1954). ***Diagnosis and Improvement of Saline and Alkaline***. Washington. U.S. Department of Agriculture.

Robinson, A.H. and R.A. Bryson, (1957). A Method for Describing Quantitatively the Correspondence of Geographical Distributions, ***Annals of the Association of American Geographers***, 47, pp. 379-391.

Rorabaugh, M.I., (1956). Prediction of Ground Water Levels on the Basis of Rainfall and Temperature Correlations, ***Transactions of American Geophysical Union***, 37 (4), pp. 436-441.

Sain, K., (1977). Development of Interstate Rivers, ***Irrigation and Power***, 34 (1), pp. 7-13.

Sankar, K., (2002). Evaluation of Groundwater Potential Zones Using Remote Sensing Data in Upper Vaigai River Basin, Tamil Nadu, India. ***Journal of the India Society of Remote Sensing***. 30 (3), pp. 119-129.

Sarma, P.B.S., (1975). Ground Water Management, ***Commerce*** 131, pp. 129-136.

Sarma, V.V.J. and A.N. Swamy, (1981). Ground Water Quality in Visakhapatnam Basin, India, ***Water, Air and Soil Pollution***, 16 (3), pp. 317-329.

Saxena, P.B., (1974). Water Resources and Potential Utilization Programme in the Suswa Valley, Dehra Dun, Garhwal Himalaya, ***Indian Journal of Regional Science***, 6 (1), pp. 39-45.

Searcy, J.K., (1959). ***Flow-Duration Curves***, U.S. Geological Survey, Water Supply Paper, 1542-A.

Sen, P.K., (1976). A Study of the Hydrological Characteristics of the Banka Basin, ***Indian Journal of Power and River Valley Development***, 1976, pp. 175-187.

Sewell, W.R.D., (1975). ***Geography of Water Resources***. Englewood, Cliffs, New York: Prentice Hall Publication Inc.

Sewell, W.R.D., (1969). Human Response to Floods, Chorley, R.J., (1969). ***Water, Earth and Man.*** London: Methuen and Co. Ltd.

Shahane, A.N., D. Thomas and P. Bock, (1976). Correlation Analysis of Hydrometeorological Data, ***Journal of Hydraulic Division, ASCE,*** 102 (10), pp. 1561-1572.

Shankarappa, B., K.S. Shamanna, and B.M. Basavaraj, (1968). Future Hydel Project on West Flowing Rivers in Mysore State, ***Indian Journal of Power and River Valley Development,*** 9 (1), pp. 27-32.

Sharan, A., (1960). ***Recent changes in Land use in the Upper Damodar Basin.*** London: University of London.

Sharma, K.P. and S.C. Jain (1978). Hydrological Investigations in Tons Basin Using Aerial Photographs, ***Irrigation and Power Journal,*** 35 (2).

Sharma, R.P., S.R. Sehgal and A.D. Gulati, (1971). Techniques and Methods for Development and Assessment of Ground Water Resources in ***Symposium on Integrated Development of Surface and Sub-surface Water Resources.*** New Delhi, Irrigation and Power Commission, pp. 31-37.

Sharma, V.K., (1974). Water Balance in the Hariyana - Punjab Plain - A Climatological Approach, ***The National Geographical Journal of India,*** 20 (3) pp. 145-159.

Sharma, V.K., (1976). Some Hydrologic Characteristics of the Damodar River, ***Geographical Review of India,*** 38 (4), pp. 330-342.

Sharma, V.K., (1978). Ground Water Potential in Hariyana, ***Geographical Review of India,*** 40 (4), pp. 301-308.

Sharma, V.K. and Jagadish Kumar, (1979). Distribution of Irrigation Water and its Impact on Agricultural Production in a Reservoir Command Area, ***Journal of Irrigation and Power,*** 59 (1), pp. 16-21.

Sharp, R.G., (1967). Estimation of Future Demands on Water Resources in Britain, ***Journal of the Water Engineer,*** 21 (3), pp. 232-249.

Shelton, M.L., (1978). Calibrations for Computing Thornthwaite's Potential Evapotranspiration in California, ***The Professional Geographer,*** 30 (4), pp. 389-396.

Singh, H., (1959). Survey of Water Potential and Overall Heeds of a Basin, ***Indian Journal of Power and River Valley Development,*** 9 (1), pp.26-32.

Singh, Minati and K. S., Yadav, (1987). Conservation and Planning of Water Resources in Chandauli - Chakia Tahsils, Varanasi. ***National Geographer***, 12 (1), pp. 51-63.

Singh, R. B., (2002). Hydrological Challenges for India. ***Annals of the National Association of Geographers. India***, 12, pp. 89-98.

Singh, S.P., (1978). The Characteristic Features of Ground Water Table of Jaunpur District U.P., ***The National Geographical Journal of India***, 24 (1-2), pp. 99-105.

Sir car, P.K., (1954). Floods and Regional Economy of West Bengal, ***The Indian Geographical Journal***, 29 (4), pp.83-88.

Sivagnanam, N. and K. Kumaraswamy, (1983). Ground Water Quality in the Vaippar Basin, South India, ***The Indian Geographical Journal***, 58 (2), pp. 127-142.

Sivanappan, R.K., (1980). Role of Water Conservation and Water Management in Rural Development, ***The New Irrigation Era***, 18 (1).

Smith, C.T., (1969). The Drainage Basin as a Historical basis for Human Activity, in Chorley, R.J., (1969) ***Water, Earth and Man***. London: Methuen and Co. Ltd.

Smith, D.I. and P. Shopp, (1999). ***The River Basin: An Introduction to the Study of Hydrology***. Cambridge: Cambridge University Press.

Smith, D.M., (1977). *Patterns in Human Geography: An Introduction to Numerical Methods*. New York: David and Charles Newton Abbot and Crane Russak and Company Inc.

Smith, K., (1983) ***Water in Britain: A Study in Applied Hydrology and Resource Geography***. London: MacMillan.

Smith, K., (1977). Water Resource Management in Scottland, ***Scottish Geographical Magazine***, 93 (2), pp. 66-79.

Smith, P.J. and R.S. Wikramaratna, (1981). A Method for Estimating Recharge and Boundary Flux from Ground Water Level Observation, ***Hydrological Sciences Bulletin***, 26 (2), pp. 113-136.

Soil Survey and Land Use Organization, (1980). ***Soil Survey Report of Periyakulam Taluk, Madurai District, Tamil Nadu***, Combatore: SSLVO (Mimeographed).

Soil Survey and Land Use Organization, (1981). ***Soil Survey Report of Uthamapalayam Taluk, Madurai District, Tamil Nadu***, Coimbatore: SSLVO (Mimeographed).

Spate, O.H.K., A.T.A. Learmonth and B.H. Farmer, (1972). ***India, Pakistan and Ceylon – The Regions.*** Bombay. Methuen and Co. Ltd., and B.I. Publications.

Srinivasan, T.M., (1968). ***Irrigation and Water Supply in South India unto 1300 A.D.*** Unpublished M.Litt. Thesis. Madras: University of Madras.

Srivastava, Pankaj K. and Bhattacharya, and K. Amit, (2000). Delineation of Ground Water Potential Zones in a Hard Rock Terrain of Bargarh District, Orissa Using IRS Data. ***Journal of the India Society of Remote Sensing***. 28 (2&3), pp. 129-140.

Strahler, A.N., (1964). Quantitative Geomorphology of Drainage Basins and Channel Networks, in Chow, V.T., (Ed.), (1964). ***Hand Book of Applied Hydrology,*** New York: McGraw-Hill Book Company.

Strahler, A.N., (1971). ***Physical Geography,*** New Delhi: Willey Eastern Private Limited.

Subrahmanyam, V.P., (1982). ***Water Balance and Its Application.*** Waltair: Andhra University.

Subrahmanyam, V.P., B. Subba Rao and Ch.V.S.N. Murthy, (1970). The Water Balance of Godavari River Basin, ***Journal of Institution of Engineers, (India),*** 50 (2), pp. 108-114.

Subrahmanyam, V.P. and C.V.S. Sastri (1971). A Climatic Appraisal of the Ground Water Potential of the Semi-Arid Region of South India, ***The Geographical Review of India,*** 33 (2), pp. 81-87.

Subrahmanyam, V.P., H.S. Rammohan and G. Parthasarathi, (1979), Appraisal of Groundwater Potential of the Dry Climates of India - A Climatic Approach, ***The National Geographical Journal of India,*** 25 (1), pp. 1-7.

Subrahmanyam, V.P., M.V. Kamaraju and M.M. Ali, (1980). Hydrometeorology and Water Balance of Krishna River Basin, ***Journal of the Institution of Engineers (India),*** 61 (CI3) pp. 146-150.

Subrahmanyam, V.P. and P. Viswanatham, (1979). A Study of Rainfall Patterns in Visakhapatnam District, ***The Indian Geographical Journal,*** 54 (2), pp. 55-64.

Subramaniam, A.R. and A. Sambasiva Rao, (1980). Potential Evapotranspiration over Maharashtra from Climatic Data using Different Methods, *Journal of the Institution of Engineers (India)*, 61 (C13), pp. 142-145.

Subramaniam, A.R. and A. Venkata Rao, (1979). Water Balance and Agriculture in the Deltaic Region of Andhra Pradesh, ***The Deccan Geographer***, 17 (3), pp. 594-615.

Subramaniam, A.R. and G.S.L.H.V. Prasada Rao, (1980), Meridianal Variation of Water Balance in Rajasthan State, ***Annals of Arid Zone***, 19 (1 and 2), pp. 51-57.

Subramaniam, A.R. and K. Uma Devi, (1979). Water Balance and Cropping Pattern of Orissa State, ***Geographical Review of India***, 41 (4), pp. 362-369.

Subramaniam, A.R. and P.V.S.S.K. Vinayak, (1982). Water Balance Studies of Trivandrum and its Vicinity, ***Transactions of the Institute of Indian Geographers***, 4 (1).

Sudhakar, M. Rao and P. Mamatha, (2004). Water Quality in Sustainable Water Management. ***Current Science***, 89, pp. 942-947

Sulkane, A. and I. Kayane, (1980). Response of Ground Water Table to Rainfall in Kumamoto Plain, ***Geographical Review of Japan***, 53 (10), pp. 666-671, (In Japanese and abstract in English).

Sukwall, B.L., (1979). River Water Management and Disputes in India, ***National Geographical Journal of India***, 14 (2), pp. 102-127.

Tamil Nadu Water Supply and Drainage Board, (1979). ***A Brochure on the Activities of TWAD Board.*** Madras: TWAD Board (Mimeographed).

Tate, D.M., (1981). Rational Water Demand Forecasting in Canada, ***Water International***, 6, pp. 130-139.

Taylor, J.A., (1970). ***The Role of Water in Agriculture.*** Oxford: Pergamon Press.

Thiessen, A.J. and J.G. Alter, (1911). Precipitation Averages for Large Areas, ***Monthly Weather Review***, 39, pp. 1082-1084.

Thirunaranan, B.M., (1938). Rivers of Palar Basin, ***The Journal of the Madras Geographical Association***, 13 (2) pp. 147-160.

Thiruvengadachari, S., (1981). Satellite Sensing of Irrigation Patterns in Semiarid Areas: An Indian Study, ***Photogrametric Engineering and Remote Sensing***, 47 (10), pp. 1493-1499.

Thiruvengadachari, S., A.K. Chakraborty and A.S. Ramamoorthy, (1978). Remote Sensing Applications in Water Resources Surveys, ***Journal of the Institution of Engineers (India)***, 58, (CI 4), pp. 165-167.

Thornthwaite, C.S. (1948). An Approach towards a Rational Classification of Climate, ***Geographical Review***, 38 (1), pp. 1955-1994.

Thornthwaite, C.W. and J.R. Mather, (1955). The Water Balance, ***Publications in Climatology***, 8 (1), Centerton New Jersey: Drexel Institute or Technology.

Thornthwaite, C.W. and J.R. Mather, (1957). Instructions and Tables for Computing Potential Evapotranspiration and the Water Balance, ***Publications in Climatology***, 10 (3). Centerton, New Jersey: Drexel Institute of Technology.

Tiwari, P.S., and R.P. Misra, (1976). Identification of Water-Sheds for Integrated Rural Development in ***Symposium on Resource Development and Planning***, November, 1976, Madras: University of Madras.

Todd, D.K., (1959). ***Groundwater Hydrology***. New York: John Willey.

Tolman, C.F., (1937). ***Ground Water***. New York: McGraw-Hill Book Company.

Trivedi, R. K., R.M. Singh and N.K. Tiwari, (2005). Hydrological studies of Ghoradongri Watershed of Betul District, Madhya Pradesh. ***Journal of the Indian Society of Remote Sensing.*** ** pp. 421-428.

UNESCO, and WMO, (1978). Assessment of Water Resource. A UNESCO/ WMO report for ECOSOC, ***Nature and Resources***, 14 (3), pp. 24-28.

UNESCO, (1978). ***World Water Balance and Water Resources of the Earth***, Leningrad: UNESCO.

UNESCO and WMO, (1981). Role and Activities of UNESCO in the Field of Water Resources, in *International Conference on Hydrology and the scientific bases for the Regional Management of Water Resources*, August, 1981, Paris: UNESCO.

United Nations, (1958). ***Integrated River Basin Development***, United Nations: Department of Economic and Social Affairs.

United Nations and ECAFE, (1955). *Multipurpose River Basin Development Part 2A. Water Resources Development in Ceylon, China, Taiwan and Phillipines*. New York:

United Nations.

United Nations and ECAFE, (1961). ***Multipurpose River Basin Development, Part 2D. Water Resources Development in Afganistan, Itan, Korea and Nepal***. Bangkok: United Nations.

United States Department of Agriculture, (1969). ***The Year Book of Agriculture, 1995: Water***. Calcutta, Oxford and IBH Publishing Company.

Unwin, D.J., (1975). ***An Introduction to Trend Surface Analysis, Concept and Techniques in Modern Geography***, No.5, Norwich, Geo Abstracts.

Utton, A.E., and L.Teclaff, (Eds.), (1978). ***Water in a Developing World: The Management of a Critical Resource***: Boulder, Colorado: West View.

Vasantha Devi, M.N., (1964). Some Aspects of the Agricultural Geography of South India, ***The Indian Geographical Journal***, 39, pp. 1-41 and 59-122.

Vedanayagam, E.G., (1977). Conservation of Surface Water and Modernization of Water Resource Development Schemes, ***Geography Teacher***, 12 (1), pp. 21-27.

Vendrov, S.L. and G.P. Kalinin, (1960). Surface Water Resources of the USSR, their Utilization and Study, ***Soviet Geographers***, 1 (6), pp. 55-60.

Venkateswaralu, A., (1959). Tunghabhadra Project: A Perspective, ***Indian Journal of Power and River Valley Development***, 9 (2), pp. 54-56.

Walling, D.E. and B.W. Webb, (1975). Spatial Variation of River Water Quality: A survey of the river Exe. ***Institute of British Geographer Transactions***, No. 65, pp. 155-171.

Walton, W.C., (1970) ***Ground Water Resource Evaluation***, New York: McGraw-Hill Book Company.

Ward, R.C., (1975). ***Principles of Hydrology***. London: McGraw-Hill Book Company.

Ward, R.C., (1978). Q Mark of MAF? Some Reflections on Hydrology today, ***Geography***, 63, pp. 301-313.

Water Supply and Drainage Board, (1984). ***Urban and Rural Water Supply in Tamil Nadu***. Madras Water Supply and Drainage Board (Mimeographed).

Weber, E.W. and M.M.H Schmidt, (1963). ***River Basin Planning in the United States***, New York: UN Conference on Science and Technology.

White, G.F., (1969). Contributions of Geographical Analysis to River Basin Development, ***The Geographical Journal***, 129, pp. 412-436.

White, G.F., (1972). River Basin Planning for Peace: The Lower Mekong, in VOA (1969), ***Geography and American Environment***, Voice of America Forum Lectures, Madras: Higginbothams.

White, G.F., (1972). Role of Geography in Water Resources Management in J.L. Douglas (Ed.) (1972), ***Man and Water the Social Sciences in Management of Water Resources.***

White, G.F., (1978). Domestic Water Supply in the Third World, ***Progress in Water Technology***, (1-2), pp. 13-19.

Wiener, A., (1972). ***The Role of Water in Development: An Analysis of Principles of Comprehensive Planning***. New York: McGraw-Hill Book Company.

World Bank, (1977). ***India: Periyar Vaigai Irrigation Projects (Tamil Nadu), Report No: 159, A, in South Asia Project Department***, Agricultural Division.

Yamomoto, S., (1971). Water Balance Study of the Groundwater on the Lower part of the River Fiji, ***Science Reports***, 10, (10), pp. 253-76.

Yamomoto, S., and Y. Musai, (1970). Rural and Urban Land Uses of the Shimizu Area in Central Japan, ***Science Reports*** (9), pp. 17-66.

Yoshino, M.M., (Ed.), (1971), ***Water Balance of Monsoon Asia.*** Tokyo: University of Tokyo Press.

Noxell, W.H., (1980). The Water Balance in Newfoundland, ***Water Resource Bulletin***, 16 (6), pp. 1030-1033.

Yu, T.Y., (1976), The Water Budget for the Nakdong River Basin, in the Republic of Korea, in Proceedings of the ***Seminar on Water Resources Management. Bangkok: United Nations Economic and Social Commission for Asia and Pacific.***

Zaltsbert, E., (1982). Methods of Forecasting and Mapping of Ground Water Tables in the USSR, ***Ground Water***, 20 (6), pp. 675-679.

Zorne, J.J. and P.H Rogers, (1978). Impact of Rainfall Enhancement on Hydrology, ***Journal of Hydraulic Division***, ASCE, 104 (5), pp. 681-694.

INDEX